Asphalt Materials - Recent Developments and New Perspective

Edited by Farzaneh Tahmoorian

Published in London, United Kingdom

Asphalt Materials - Recent Developments and New Perspective
http://dx.doi.org/10.5772/intechopen.111354
Edited by Farzaneh Tahmoorian

Contributors
Amrita Roy, Berthold Best, Bharat Kapgate, Cristina Oreto, Eche S. Okem, Francesca Russo, Gianluca Dell'Acqua, Homa Kohzadi, Jung Huh, Kasilingam Rajkumar, Mohamed M.H. Mostafa, Mohammad Soleiman-Beigi, Muhammad Aniq Gul, Muhammad Mudassir Ahmad Alwi, Nathan N. Chilukwa, Rosa Veropalumbo, Saeed Toolabi, Simone Malvezzi

First published in London, United Kingdom, 2025 by IntechOpen
IntechOpen is the global imprint of INTECHOPEN LIMITED, registered in England and Wales, registration number: 11086078, 167-169 Great Portland Street, London, W1W 5PF, United Kingdom

For EU product safety concerns: IN TECH d.o.o., Prolaz Marije Krucifikse Kozulić 3, 51000 Rijeka, Croatia, info@intechopen.com or visit our website at intechopen.com.

British Library Cataloguing-in-Publication Data
A catalogue record for this book is available from the British Library

Asphalt Materials - Recent Developments and New Perspective
Edited by Farzaneh Tahmoorian
p. cm.
Print ISBN 978-0-85466-377-4
Online ISBN 978-0-85466-376-7
eBook (PDF) ISBN 978-0-85466-378-1

If disposing of this product, please recycle the paper responsibly.

Meet the editor

Dr. Farzaneh Tahmoorian is a Senior Lecturer at Central Queensland University (CQU) in Australia, with over a decade of experience in construction and related fields. She has contributed to numerous projects involving road construction, landfill development, and waste management. Farzaneh holds an M. Sc. from the University of Technology Sydney (UTS) and a Ph.D. from Western Sydney University. Dr. Tahmoorian's research focuses on incorporating waste materials into asphalt, concrete, and other pavement materials. She is the recipient of the Australian Postgraduate Award and the Western Sydney Top-up Award. As a researcher, Dr. Tahmoorian has published over 25 peer-reviewed journal and conference papers on asphalt mixes and sustainable pavement technologies. Her expertise spans a broad range of topics, including asphalt mix design, binder modification, concrete pavements, ground improvement techniques for roads and railways, pavement engineering, sustainability, circular economy, life cycle analysis, and waste management. Dr. Tahmoorian's extensive professional experience and research achievements underscore her dedication to civil engineering and her commitment to advancing sustainable infrastructure solutions for future generations.

Contents

Preface **IX**

Chapter 1 **1**
Innovative Approach to Road Infrastructure Asset Management
by Berthold Best

Chapter 2 **23**
Crumb Rubber Modified Asphalt: Fundamentals to Recent Developments
by Amrita Roy, Kasilingam Rajkumar and Bharat Kapgate

Chapter 3 **39**
Strengthening Substandard Road Materials with Nanoemulsion-Based Stabilisation: An Experimental Study
by Eche S. Okem, Nathan N. Chilukwa and Mohamed M.H. Mostafa

Chapter 4 **55**
A Comprehensive Scientific Theory for Hot Mix Asphalt Design Aiming Accuracy and Simplicity for Asphalt Pavement Construction and Mix Quality Check
by Jung Huh

Chapter 5 **73**
Self-Healing Asphalt Using Steel Fibers: Induction Heating
by Muhammad Aniq Gul and Muhammad Mudassir Ahmad Alwi

Chapter 6 **105**
Natural Asphalt: A Novel Bitumen Modifier and Asphalt Filler
by Mohammad Soleiman-Beigi, Homa Kohzadi and Saeed Toolabi

Chapter 7 **135**
Road Asphalt Pavements Made by Bottom Ashes from Waste-to-Energy Plant as a Filler
by Rosa Veropalumbo, Simone Malvezzi, Cristina Oreto, Francesca Russo and Gianluca Dell'Acqua

Preface

This edited volume is a collection of peer-reviewed and relevant research chapters concerning developments in the field of asphalt materials. The book includes scholarly contributions by various authors and is edited by an expert in the field of asphalt materials. Each contribution is presented as a standalone chapter, complete in itself yet directly related to the book's overarching topics and objectives.

The target audience comprises scholars and specialists in the field.

IntechOpen

Chapter 1

Innovative Approach to Road Infrastructure Asset Management

Berthold Best

Abstract

Efficient management of municipal road assets demands consistent condition monitoring. Traditional methods, encompassing measurement techniques and visual inspections, fall short in the face of urban environment dynamics, requiring substantial human and financial resources. The periodic assessments currently in use make it challenging to maintain an up-to-date perspective on road conditions and provide continuous road availability. In response to these challenges, one noteworthy outcome of the current research is the introduction of vibration monitoring. Leveraging advanced data acquisition techniques and Artificial Intelligence (AI) in the data evaluation, this approach enables continuous data collection and condition assessment with minimal human intervention. Allowing for real-time updates facilitates early detection of damage and supports the development of optimal maintenance strategies, mitigating the need for costly premature resurfacing or rehabilitation. Its impact extends beyond facilitating faster and more reliable decision making in finance, operations, and maintenance. It is crucial to outline quantifiable benefits resulting from this approach, such as improved road sustainability, increased availability, and potential cost savings. Despite its potential, the adoption of vibration monitoring for road condition assessment may encounter challenges, such as implementation costs, technological barriers, regulatory compliance, and a lack of scalability to varying scenarios. Acknowledging these hurdles is crucial for a comprehensive understanding of the proposed solution. Collaboration with stakeholders, including local communities and relevant authorities, is essential for successful implementation. In conclusion, the introduction of vibration monitoring marks a significant leap forward in the continuous assessment of road conditions. It offers tangible benefits such as improved sustainability, increased road availability, and potential cost savings. Acknowledging challenges, connecting to existing research, and addressing practical implementation considerations are essential steps toward realising the full potential of this innovative approach to municipal road asset management.

Keywords: AI, asset management, building information modelling, BIM, degradation, monitoring, vibration, maintenance, data collection, road pavement, international roughness index, IRI

1. Introduction

Every road infrastructure object wears out through climatic conditions and traffic loads. Intensive use of road pavements, irregular maintenance, and inappropriate treatment methods cause damage to the pavement surface and deterioration of the pavement condition. As a result, the quality of the ride and the efficiency of the road availability are reduced. However, pavement management systems focus on accurate performance assessment, which requires accuracy and a reliable monitoring system of the pavement condition.

There are three main methods of collecting information on road conditions. The most common method, especially on trunk roads, is metrological condition monitoring. High-speed road monitors, consisting of laser scanners combined with high-resolution line cameras, record data on the longitudinal and transverse evenness and substance characteristics of the surface. These high-speed measurement systems can be used without disturbing the flow of traffic.

Visual image-based condition monitoring, the second most widely used method of road condition monitoring, is based on the visual analysis of photographic and video images and allows both asset and condition data to be recorded at the same time. This advantage requires digital stereo image cameras and measuring cameras directed at the road surface. Infrastructure objects and their physical characteristics are measured using stereoscopic measurement techniques and 3D laser scanning.

The oldest and most proven method for road condition assessment is visual assessment. This type of recording is only suitable for minor roads and/or side areas like bicycle paths in order to exclude traffic hazards for the monitoring personnel. The use of mobile digital logging devices (e.g. tablet PCs) is widespread to minimise mistakes in the recording. The road network is stored on the device as a digital network model. GPS support allows direct geo-referenced input of the recorded condition features. To facilitate recording, the surface types (road, footpath, cycle lane, etc.) and surrounding buildings are visualised for the operator. With this recording method, the design of the input interface influences the quality and speed of condition monitoring. Progress is comparatively slow (an average expert evaluates 3–4 km per hour).

The three methods described here have the disadvantage that they can only be carried out periodically, on average every 4–5 years, due to the financial and personnel costs involved. This also means that the state of the infrastructure is only glimpsed. Changes between survey campaigns due to traffic, climate, possible excavations, or other adverse changes to the road are not recorded. There is therefore a need for continuous road condition monitoring. However, it must be simple, not require a lot of manpower and, above all, be cost-effective. Road condition monitoring using vibration sensors appears to meet these requirements. The following considerations are therefore intended to shed more light on this method in principle, to examine existing approaches for their suitability and to compare the results with conventional condition monitoring. The aim is to enable road managers to assess whether vibration monitoring is suitable for their purposes.

2. Innovations in road condition monitoring

2.1 Condition monitoring

In general, there are three groups of condition characteristics, which are obtained by different technologies:

Roughness and rutting are characteristics of the evenness in the longitudinal and transverse directions. A common method of measuring roughness is multiple scanning of the road surface using laser triangulation sensors mounted vertically downwards on a rigid beam of 2 or 4 m in length. This is vibration-isolated and mounted under the vehicle. The longitudinal profile of the road is determined from the laser measurements, which are typically taken at 0.1 m intervals.

Laser triangulation sensors mounted on a rigid frame at the rear of a carrier vehicle are used to detect rutting, transverse evenness, and road slope. This rigid frame is also vibration decoupled. A width of up to 4 m can be measured in one pass with the respective outer sensors on the measuring beam set at an angle.

Skid resistance is measured using a SCRIM (Sideway-force Coefficient Routine Investigation Machine). This is metrological equipment mounted on a truck to measure the wet skid resistance of a road surface using the sideways force principle. Skid resistance is less important on urban roads, where traffic speeds are lower. It is therefore not included in the regular condition monitoring campaigns in municipalities.

Pavement texture features are detected by high-resolution line scan cameras. These are usually mounted to the left and right of the rear of the monitoring vehicle and record the road surface synchronously. This is done by illuminating the road with LED lighting. This produces a video of the road surface in 256 grey scales. The surface images are compressed at 10 m intervals. In addition to these images, the front cameras capture a colour image of the vehicle roof for reference, synchronised with the road surface image. The condition of the surface (cracks, patches, and other surface defects) is determined from these video images.

2.2 Condition assessment

Condition monitoring provides measurements with physical units. For the purpose of comparison at object and network level, it is necessary to convert the condition variables into dimensionless condition values. A common unified grading scale ranges from 1 to 5, with 1 being the best and 5 being the worst.

The objectives of the condition assessment are as follows:

- Scoring: The objective state variables are technically evaluated.
- Comparability: By converting to uniformly scaled condition values, different condition characteristics and different traffic area types can be compared.
- Linkage: The uniformly scaled state values can be linked to complex state values (partial target values, total value).
- Processing: Condition values are well suited and intuitively understandable for display on maps, for statistical evaluations and for the monetary rating of infrastructure assets.

For the purpose of conversion, defined normalisation functions are applied to the condition variables. These functions vary from country to country and even within countries for the application to different types of road networks. The higher the safety and comfort requirements, the more stringent the conversion functions. Rutting on urban roads does not have the same serious consequences as on motorways due to the lower speeds within urban areas. Therefore, a deeper rut on an urban road may result in better condition scores than a comparatively shallower rut on a motorway.

2.3 Systematic road maintenance

A management cycle, also known as a continuous improvement cycle, plays a key role in the systematic approach to road maintenance. This cycle enhances the efficiency and effectiveness of the entire maintenance process by following six distinct stages: road condition monitoring and evaluation, planning, coordination, execution, measure monitoring, and evaluation (see **Figure 1**). Each stage contributes to a well-structured and dynamic road maintenance strategy, ultimately leading to improved road safety, prolonged infrastructure lifespan, and enhanced overall transportation efficiency for the community.

- Road Condition Monitoring/Assessment, Evaluation and Issue Identification: This phase includes the collection of road condition characteristics and the evaluation of condition data. Road safety and the identification of safety issues complete this phase as input for the planning phase.

- Planning of Maintenance Measures: During the planning phase, road authorities carry out a comprehensive assessment of the condition data of the roads within their jurisdiction. This assessment provides the basis for identifying and prioritising maintenance needs. By carefully analysing road conditions and taking

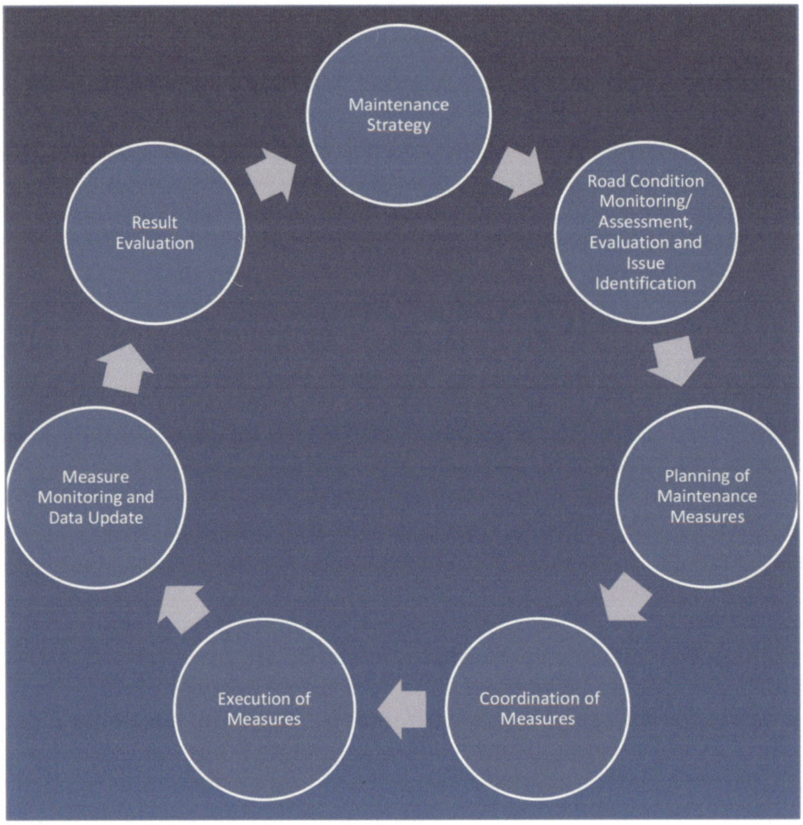

Figure 1.
Management cycle for systematic road maintenance.

into account factors such as traffic patterns, weather impacts and intensity of use, authorities can strategically allocate resources to areas that require immediate attention. Effective planning ensures that maintenance efforts are focused on critical areas, optimising the use of resources.

- Coordination of Measures: All maintenance activities are coordinated locally and in a timely manner within the network. This ensures smooth traffic flow within the network and minimises detours. Combining maintenance activities can save money and reduce traffic disruption. Planned excavations for the maintenance of sewers, pipes and cables must be considered.

- Execution of Measures: In the execution stage, maintenance activities are carried out based on the established plan. This includes carrying out repairs, implementing preventive measures and addressing identified maintenance needs. Timely implementation of maintenance activities is essential to prevent further deterioration of the road infrastructure. By adhering to the planned maintenance schedule, road authorities can minimise disruption to traffic flow and ensure the safety of road users.

- Measure Monitoring and Data Update: The monitoring phase involves continuous tracking of the progress of maintenance activities. This stage allows the road authority to monitor the implementation of the maintenance plan and identify any deviations from the intended course. Regular monitoring provides an opportunity to identify potential problems or challenges that may arise during the execution of maintenance tasks. By staying vigilant and responding to real-time data and feedback, authorities can take corrective action as needed, ensuring that maintenance activities stay on track and effectively address road condition issues.

- Result Evaluation: The evaluation stage is crucial for assessing the results of maintenance activities. Road authorities analyse the results achieved through the implementation of the maintenance plan. This evaluation helps to determine the impact of maintenance activities on road condition, safety and overall infrastructure integrity. By identifying successes and areas for improvement, road authorities can refine their maintenance strategies for future cycles. The evaluation process provides valuable insights to inform decision making and guide adjustments to maintenance plans, resource allocation and execution methods.

By embracing and consistently implementing this management cycle, road maintenance evolves into a structured, adaptive process. This approach not only ensures the longevity of road infrastructure but also enhances road safety and contributes to the overall efficiency of transportation systems. The continuous improvement cycle provides road authorities with a systematic framework to proactively address road maintenance needs, effectively manage resources, and create sustainable road networks that cater to the needs of the community.

2.4 Basics of accelerometer technology

Acceleration sensor technology is increasingly being used in innovative approaches to road condition monitoring. This involves mostly measuring the vertical acceleration of a vehicle as it travels. In some cases, the use of smartphones or other consumer electronic devices is being offered as an effective, cheap and easy-to-handle means for data collection [1–3].

Today's smartphones tend to have a wide range of sensors built into them. Accelerometers, for example, are used in smartphones to execute commands by shaking the smartphone, to record the number of steps taken with the pedometer, or to play various games. A gyroscope is used to automatically align the display when the smartphone is rotated, and a GPS module is used to determine the location in navigation applications. They therefore appear to be well suited to this condition monitoring method.

After all, smartphones are not primarily designed as measurement systems. This is reflected in the fact that little or no information is available about the sensors installed, so their quality depends on the smartphone manufacturer or even the model. In addition, a smartphone tends to become obsolete and needs to be replaced more quickly than a purpose-built metrological unit. Apps can also be changed at any time without users being informed. Thus, according to various studies, the hardware and software built into a smartphone are capable of functioning as a measurement system, but it is not designed to do so.

It remains to be investigated what accuracy and repeatability can be achieved with such vibration sensors for recording road condition characteristics relevant to local authorities. The first priority is to select suitable and easy-to-use vibration sensors.

A prerequisite for road condition assessment using accelerometer technology is a correlation between the measured vertical acceleration and the condition of the road surface. International studies have already investigated accelerometer measurement setups based on the International Roughness Index (IRI). First mentioned in a 1986 World Bank publication, the IRI is used in many countries around the world, and its collection is still required by the World Bank as a regular condition monitoring tool for road projects financed by the Bank. Therefore, it is obvious that acceleration data obtained by vibration sensors can be related to the IRI scale.

The IRI is used to describe the effects of roughness on a vehicle. Thus, this index can be used to make statements about the following:

- total vehicle operation costs
- general ride comfort
- dynamic wheel loads
- general surface condition

However, the description of the roughness is not sufficient to describe the total road condition. Further investigations have shown that not all pavement characteristics can be captured by vibration monitoring.

Accordingly, the following condition characteristics can be recorded by means of vibration sensors:

- short-wave roughness in the longitudinal profile,
- fretting,

- transverse and crocodile cracks,
- open seams and joints,
- patches.

However, it is not possible to record.

- long-wave roughness,
- longitudinal cracks,
- evenness in the transverse profile, especially rutting,
- binder accumulations.

The IRI is defined as a property of a single wheel track profile because many measurement systems for profile detection can only record a single longitudinal profile. Basically, a measurement interval of 300 mm is required for the IRI; 600 mm is sufficient for less accurate results. For roads that have very good evenness, an interval of less than 300 mm is required. The recorded longitudinal profile should have a constant slope between recorded high points, resulting in the recommended unit of IRI of metres per kilometre (m/km), millimetres per metre (mm/m), or slope × 1000 (**Figure 2**).

2.5 Research project

The aim of the research project is to develop a cost-effective measurement system for continuous monitoring of road conditions in urban areas. The use of new data

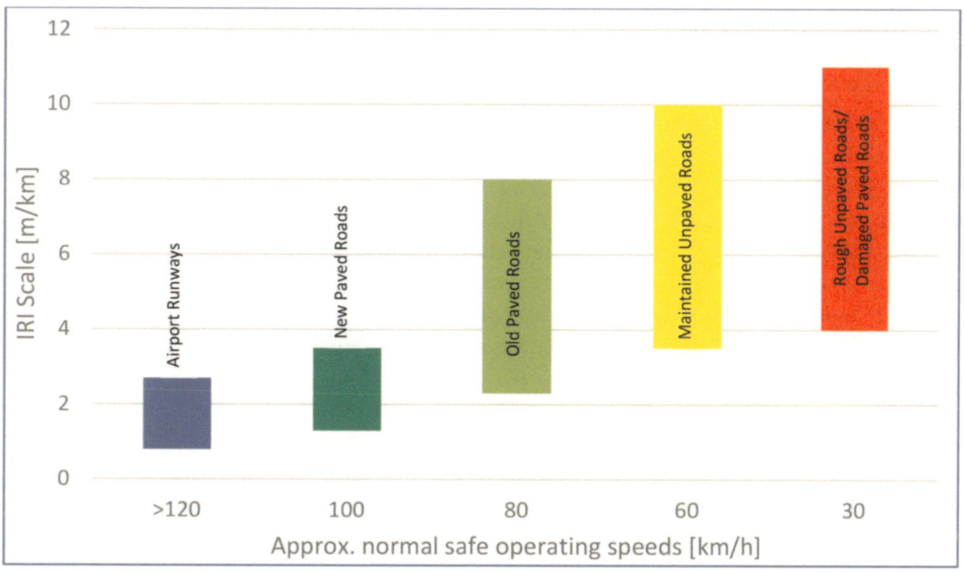

Figure 2.
IRI range by road classes and speeds.

collection methods should enable continuous data monitoring and condition assessment with a minimum of manpower. This will support the maintenance management of municipalities and has an impact on economic and political concerns.

Initially, the research started with a smartphone attached to a bicycle [4]. The vibrations during the ride were measured simultaneously by two smartphones. Smartphone 1 was mounted on the right side of the handlebar, Smartphone 2 on the upper bar of the frame. Both were aligned horizontally in order to assign the directions of the acceleration pulses during evaluation, with the y-axis in the direction of travel and the z-axis vertical. Smartphone 1 used the app "SensorLog" and Smartphone 2 "Phyphox". "Sensor Log" allows to log and stream the smartphone's sensor data. "Phyphox", which stands for "physical phone experiments", is an app developed by Germany's RWTH Aachen University. It is primarily designed to experiment with the phone's own sensors. Additionally, for reference purposes, an action camera, "GoPro Fusion", was mounted on the handlebars with its own bracket and roughly aimed at a point in front of the front tyre with a lens in the direction of travel.

The evaluation of the tests carried out showed that the camera used is suitable for recording the road surface at speeds of up to 50 km/h with sufficient quality. In addition, the evaluation of the collected vibration data can be assigned to a specific road condition with acceptable accuracy.

Since higher and more steady speeds as well as longer measurement runs, are possible with a passenger car, the focus of the investigations was then placed on this method. In addition, the roads to be investigated are mainly used by passenger cars, so the university's own Volkswagen (VW) bus model, "T5", was used.

Instead of Smartphone 2, a USB acceleration sensor from "Code Mercenaries", the model "JoyWarrior 24F14-WP" was fitted in the car, the measured values of which were recorded with a laptop in the passenger compartment. This was used to record accelerations at 20 Hz and the roll, pitch and yaw angles of the sensor. The "GoPro" was attached by means of a suction cup in the centre of the front area of the bonnet at a distance from the road surface of about 1.5 m.

The vibration data was recorded with the smartphone, the app "SensorLog", and the USB accelerometer. The smartphone was mounted horizontally in the glove compartment on the passenger side. The USB sensor was placed between the front seats on the floor of the interior, secured against slipping by an adhesive strip and connected to a laptop. Both devices were aligned so that the y-axis was pointing in the direction of travel and the z-axis was pointing vertically upwards.

The result of this condition monitoring test showed that only condition features affecting the roughness can be recorded. Structures embedded in the road surface, such as manhole covers, can also be detected. Features of the transverse unevenness, such as ruts or longitudinal cracks that run along the direction of travel, cannot be recorded. Since the driving speed has an influence on the level of the measured values, this should be kept as constant as possible during the measurements and documented continuously.

In the course of the tests carried out, it was investigated whether the action camera used delivers sufficient video quality to detect surface damage with sufficient reliability at usual driving speeds of up to 50 km/h. For a positional allocation of the image data, the GPS data stored by the camera was transferred into the videos and imported into a GIS as a shape file. This allows us to determine the position of any damage visible in the footage and to record the distance travelled during the measurement. The acceleration measurement data collected by means of a smartphone sensor was automatically displayed graphically by a macro and synchronised with the

DOI: http://dx.doi.org/10.5772/intechopen.1004580

recording time of the videos. For the evaluation of the results, the processed image recordings and vibration data were compared, and the recognisability of individual condition features was documented. Deflections in the vibration data above a selected threshold could be associated with existing surface damage with sufficient reliability. This concluded the first campaign within the project. In particular, there were difficulties in synchronising the vibration and image data. In addition, it was not possible to unambiguously associate the vibration data with specific damage, but only to determine that there was specific damage to the road surface.

Lessons learned from the first step of the project were that the road condition detection method requires GPS, video and acceleration data to be captured while driving. GPS and video data were previously recorded with an action camera. Acceleration data was initially collected by smartphones and then by USB acceleration sensors. Preliminary evaluations showed that the action camera not only records GPS and video data but also acceleration data. When all three types of data are recorded by one device, synchronisation of the data is simplified. For this reason, acceleration measurements were carried out with the "GoPro HERO9 Black" in further development, and the suitability of this action camera as a measurement system was tested [5–7].

Three measurement runs were carried out with a VW T5 and the "GoPro HERO9 Black". See **Figure 3** for a video clip of the damage at the point of maximum acceleration deflection. The area with the maximum measurement deflection is recorded in the following diagram (see **Figure 4**).

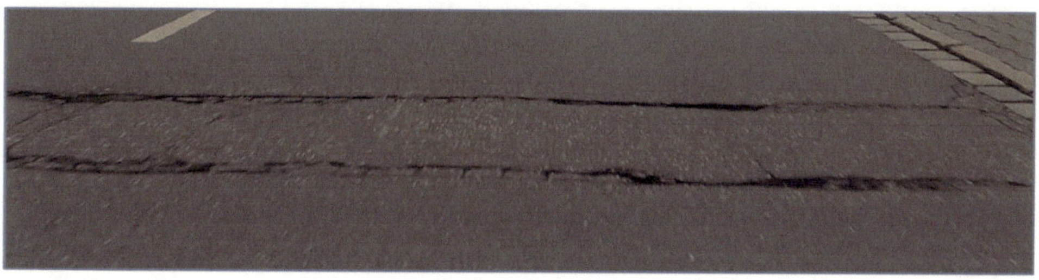

Figure 3.
Road damage at maximum acceleration deflection [5].

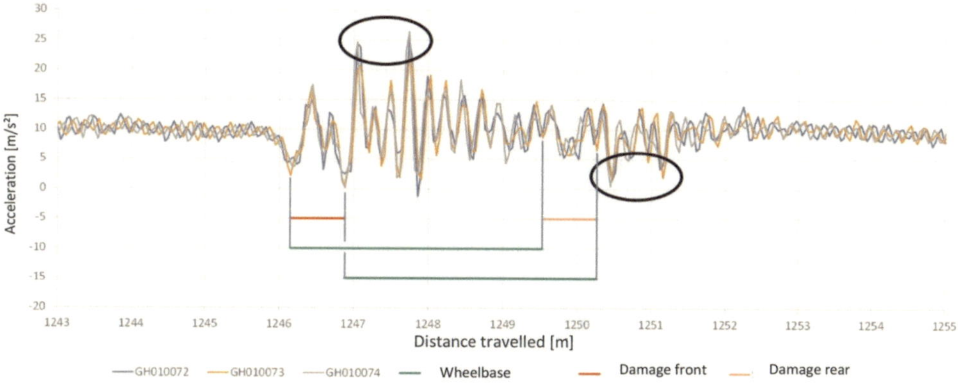

Figure 4.
Acceleration values [5].

The graph shows the acceleration values on the y-axis and the distance travelled in metres on the x-axis (see **Figure 4**). It shows the course of the acceleration data for all three measurement runs. At the beginning, the acceleration values oscillate around the rest value of 9.81 m/s^2. The curve is then stimulated by the road damage and reaches a low point at 1246.15 m in data series GH010073. The next similarly pronounced low point is at 1246.88 m. The distance between the two points is 0.73 m. This corresponds approximately to the longitudinal extent of the damage. The previous oscillations were only caused by the two front wheels driving over the damage. So, there are two more extremes. One low point is at 1249.52 m, and the other at 1250.26 m. The distance between these two points is 0.74 m. The distance from the first low point of the front wheels to the first low point of the rear wheels is 3.37 metres, and the distance from the second low point of the front wheels to the second low point of the rear wheels is 3.38 metres which equals the wheelbase of a VW T5 (exactly: 3.4 m). It should be critically noted that the effects of the rear wheels on the course of the accelerations are not apparent in the diagram without additional knowledge. The small effects can be attributed to the large distance between the GoPro and the rear wheels. It is notable that the accelerations after passing over the damage overlap in such a way as to produce clearer extremes. These extremes are circled in black in the diagram in **Figure 4**. The distances between the extremes also reflect the extent of the damage in the longitudinal direction and the wheelbase of the measurement vehicle. These superimposed extremes are clearly visible in the diagram and provide an initial basis for identifying the damage from the acceleration data and determining its extent in the direction of travel. In the area shown, all three series of measurements were taken at a constant speed. As the series of measurements progressed, there were speed differences between the series, and it became obvious that the longitudinal acceleration of the vehicle had an effect on the vertical acceleration. This phenomenon is known in vehicle engineering as “pitching”.

Overall, the GoPro HERO9 Black is suitable as a measurement system and can certainly be used for further research. However, it has been shown that it would be beneficial to have more information and settings available on the accelerometer. As the GoPro does not provide any information about the installed components and no settings can be made on the accelerometer, a separate measurement system with external sensors was set up in the next phase of development.

A single-board Raspberry Pi 3 was used to control the measurement system setup for the next phase of the project, controlling two accelerometers and a Global Navigation Satellite System (GNSS) module with Python code. Unlike the smartphone and the GoPro, the chosen sensor cannot receive a position signal. There is no prepared interface to couple the sensor to a GNSS module. If both the sensor and the GNSS module are controlled by the same single-board computer, all measurements can be stored with the computer’s time and synchronised over time. The measurement system is currently being used on the university’s VW T5. In order to be able to use the measurement system on different vehicle models in the future, vehicle-specific influencing factors must be minimised. Therefore, these must first be defined. A MATLAB/Simulink model has been developed to simulate a quarter vehicle.

It is a one-dimensional model with two degrees of freedom, which allows the vertical movement of a vehicle to be represented. When comparing the simulation results with the measurement results from test runs, it was discovered that the model can represent the vertical acceleration at the wheel carrier better than the vertical acceleration of the body. This is due to the fact that the entire vehicle, with its

Figure 5.
Sensor under wheel carrier (left) and sensor in vehicle interior (right) [5].

vehicle-specific components, has an influence on the vertical acceleration during the measurements of the body.

In order to find the vehicle-specific properties of the VW T5, a measurement setup was designed. For the measurement runs, an acceleration sensor was attached to the wheel carrier and a second sensor in the vehicle interior (**Figure 5**).

The sensors were installed on top of each other as far as possible so that the measurement results are not influenced by the position. Different results in the measurements can only be attributed to the height difference of the sensors and, above all, to the vehicle components involved.

Before a new measurement system, for example, the one with the Raspberry Pi can be used in road condition assessment, its suitability must be tested.

2.6 Validation process

During the validation process, the relationship between the acceleration data and the state of the road must be verified. Currently, the condition is documented and subsequently evaluated as part of a metrological road condition survey using high-speed certified systems. With the help of this evaluation, which is obtained through established and standardised recording methods, the validation of the alternative method can be undertaken.

In contrast to many comparable studies, the acceleration data with the method developed at Ohm University is not recorded by smartphones. Due to the different components installed in smartphones, no comparable measurement can be made. In addition, smartphones are located in the interior of a vehicle and are influenced by the vehicle-specific influencing factors of the vehicle. In this project, external, commercially available acceleration sensors are used, which are installed at defined points on the vehicle body. This allows measurements to be taken over a longer period of time under the same conditions. Vehicle-specific factors such as natural frequencies and damping are filtered out by prior calibration measurements, giving the measured values a higher degree of accuracy.

The current system of systematic and periodic road condition monitoring is expected to become much simpler and cheaper with the use of vibration sensors. In addition, permanent condition monitoring will take place when this technology is installed in conventional vehicles that regularly operate in the area of a road network. The initial idea is to use vehicles from the Road Administration, but buses and taxis could be used as well. With permanent condition recording, a continuous inventory of the asset “road” would take place. With this current method of recording the condition of roads, the most common inventory at a key date can be replaced by a permanent inventory. The main advantages are that the inventory is spread over the year, and there is no time pressure to repeat the key date assessment. In addition, the inventory can be carried out at a time when staff are not tied up with other seasonal priorities, such as carrying out construction work in the months when the weather is likely to be favourable.

The longer-term goal is to link sensors as acquisition modules with evaluation modules and to use artificial intelligence to evaluate the measurement results and make them available to the user as a single module.

Up to now, the research project has proven the basic feasibility of road condition monitoring using vibration sensors. In the next steps of research, the objectivity of the method, as well as its reliability and validity, must be proven. This procedure has to follow these six steps:

Step 1: Collection of measurements under known conditions.

Known conditions in this case mean defined, measurable scenarios that generate the acceleration data.

To provide a benchmark for comparison and calibration for the following steps, the first step is to carry out measurement runs with regard to the influence of speed, rotational speed, transmission and inherent vibration from the engine of the vehicle, as well as the roughness of the road surface on the vibrations generated. The measurement drive over defined known obstacles is also included in this step. The measurement runs must be carried out with an optimised measurement setup and accurate but commercially available measurement equipment.

Step 2: Evaluation of the influences.

In the second step, the measurement data from step 1 has to be evaluated using various methods, such as Fast Fourier Transformation. The aim is to calculate the influence of speed, rotational speed, gear and inherent vibration of the engine.

At the same time, a MATLAB simulation must be created for the test vehicle. After entering the damage dimensions and the vehicle-specific properties, the generated vibrations are collected as output. These must then be compared with the data collected with the results of a metrological road condition survey for their evaluation.

Step 3: Measurement runs for extensive data set.

The recording of measured values on a road by means of vibration sensor technology is carried out. Each condition feature must be present several times in different forms in the collected data. These measurements form the data set for the artificial intelligence evaluation in step 5.

Step 4: Validation of the measurement data.

In step 4, the measurement data from step 3 has to be validated on the basis of the current road condition survey with conventional metrological techniques. Limit values have to be calibrated and defined so that the threshold values for the condition of the transverse and longitudinal evenness and for surface damage specified in the applicable regulations can be transferred to the vibration recording. The correlation of the parameters is the decisive quality criterion for the alternative method of condition detection by means of vibration sensors.

Step 5: Automated evaluation and visualisation.

This step involves the construction of a prototype as a test model with unrestricted technical functionality in relation to the final product, using commercially available suitable vibration sensors and a GNSS module controlled by a single-board computer installed in a housing with multifunctional mounting options on various parts of the vehicle body and live data transmission to the cloud using 5G/4G technology. The automated, AI-based analysis of data in post-processing and visualisation of results in a GIS is the result of the test runs. Calculation runs are carried out using the current condition values to optimise the maintenance strategies in order to minimise the construction costs and construction time of maintenance measures.

Step 6: Use of the prototype.

Use of the prototype in a selected municipal area in operation.

2.7 Use case

As part of the project, a use case with measurement runs was established on a selected test track. The criteria are as follows:

- stretched track layout,
- at least one pavement change,
- different existing road surface conditions,
- maximum speed of 50 km/h, and
- existing comparative data from metrological road condition monitoring.

To ensure comparability and calibration of the condition data obtained from the vibration sensors, the roughness was first measured using a measuring system with the approval of the German Federal Highway Research Institute. The roughness measurement was carried out using multiple laser scanning. There are several ways to assess the roughness of a road. The most common ones in Germany are the "General Unevenness" measured in mm over a base of 4 m and the "Weighed Longitudinal Profile". The IRI is not common in Germany but can be calculated from the results of multiple laser scanning.

The alternative measurement setup using an acceleration sensor was installed on the passenger side wishbone of a VW T5. The passenger side was chosen because the laser scanning of the measurement vehicle for the calibration measurement was also carried out on the right side of the vehicle. The GNSS sensor was mounted on the dashboard directly behind the windscreen in the centre of the vehicle. As with the previous test drives, the sensors were controlled by a Raspberry PI [8].

The acceleration data obtained was calculated and synchronised with the GPS data. A macro was used to calculate the data in Excel. After this synchronisation, the measurement results were displayed in a geo-information system.

Examination of the recorded acceleration values showed good agreement with the local condition characteristics of transverse cracks, material loss (potholes) and patching [9]. In addition, the accelerograms showed changes in the road surface, manhole covers and valve caps. It is not possible to distinguish the recorded vibrations caused by pavement elements from those caused by surface condition features without accompanying video analysis.

In order to compare the conventional measurement with laser sensors and the alternative method using vibration sensors, the IRI was used, which can be calculated from the data of the laser sensors. A relationship was then established between the IRI and the Root Mean Square (RMS) of the vertical acceleration gathered by the vibration sensors. The RMS value provides a representative description of the "average size" of a data set or function, considering both positive and negative values. This approach is particularly useful for periodic signals. It is used in various contexts, including signal processing and statistics, to quantify the dispersion of data points around the mean value. For the IRI and RMS to be comparable, the vertical acceleration sections must be divided in such a way that the IRI and RMS are calculated in the same way. For this use case, 100-metre sections have been created.

The RMS is calculated using the following equation:

$$\mathrm{RMS} = \sqrt{\frac{1}{N}\sum_{i=1}^{N}(a_{zi} - g)^2} \tag{1}$$

where N is the number of acceleration recordings in the section, a_{zi} is the maximum value of acceleration in section i and g is the gravitational acceleration (9.81 m/s^2). The IRI values and the RMS values were calculated for nine selected sections of the test track and compared (see **Figure 6**).

The correlation between IRI and RMS derived from the nine test sections has a coefficient of determination of 0.932 and a correlation coefficient of R = 0.965. **Figure 7** compares the IRI and RMS values calculated for each section.

With the help of this correlation between IRI and RMS, a way was sought to calculate the IRI from the acceleration data alone and thus to be able to derive a conventionally introduced state variable, the IRI, from the measurement results of the vibration sensors. In the present case, with the measurement setup described, a driving speed of approx. 50 km/h and a 100 m long section were used:

$$\mathrm{IRI} = 0,8278 \text{ x } \mathrm{RMS} + 0,5415 \tag{2}$$

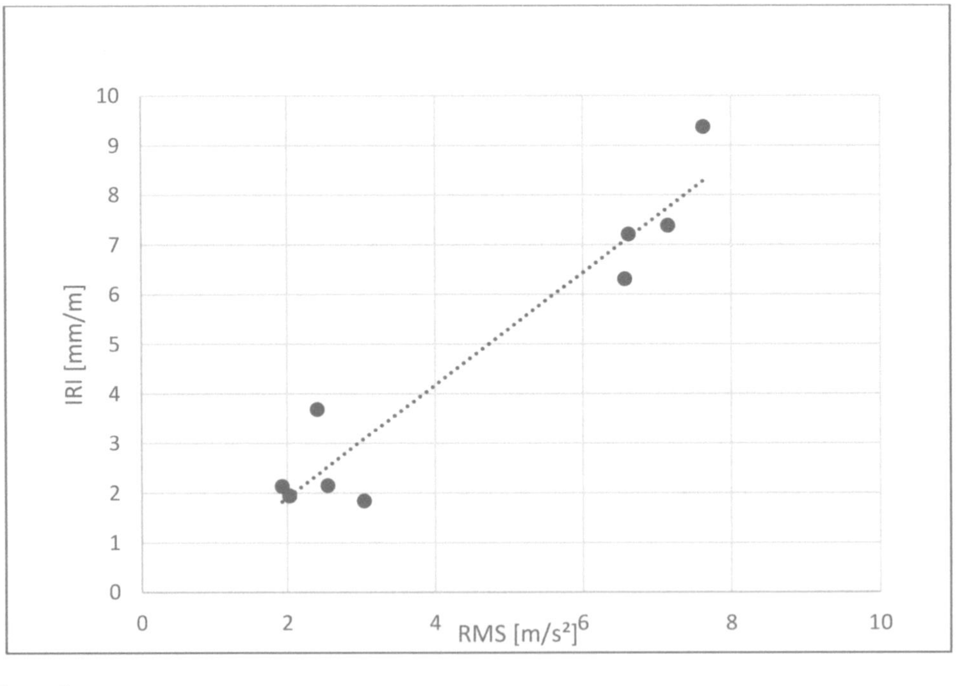

Figure 6.
Correlation between IRI and RMS.

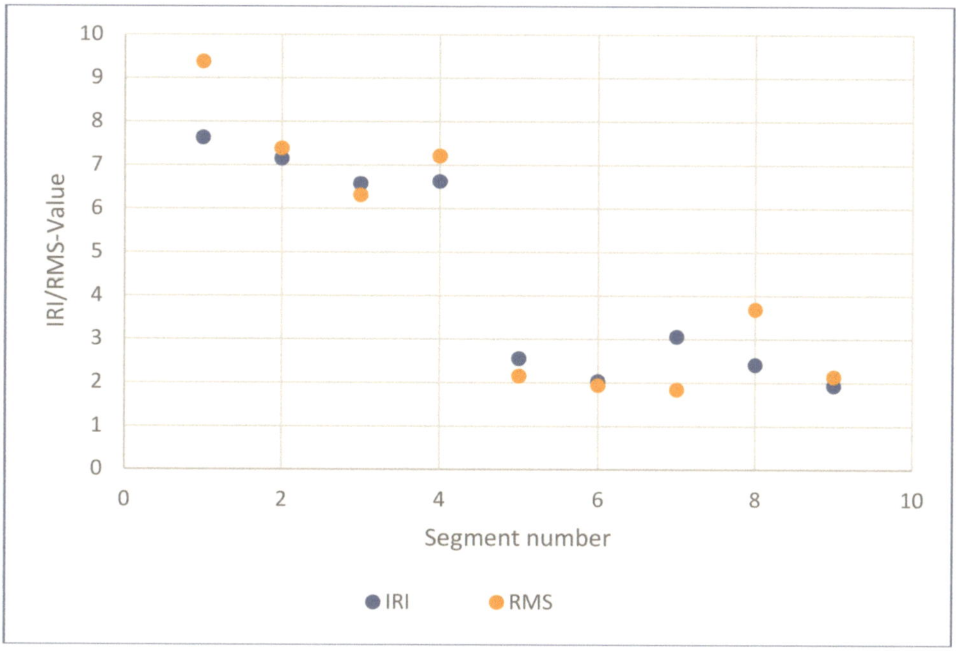

Figure 7.
Relation between IRI and RMS.

A good correlation between the IRI and the RMS of the vertical acceleration was made possible by the chosen location of the accelerometer mounting at the bottom right of the measurement vehicle. Due to the proven high correlation, the IRI can be estimated using the alternative measurement setup if required, which would be more cost-effective than purchasing a pavement measurement device capable of providing the IRI. If in the future, the measurement setup and speed of approximately 50 km/h is used as in this study, the IRI for a 100 m section can be approximated using Eq. (2).

The evaluation of the RMS is imprecise because the ride quality is only described by the factor "convenience". Several influencing factors would describe the RMS more accurately, as the comfort of a ride is a personal perception and can vary. The main criticism of both the IRI and the alternative measurement setup is that both only measure the roughness in the wheel track. Photographs and video footage are used to assist in the assessment as they allow the damage to be allocated in a meaningful and concrete way. To cover more of the road surface with acceleration sensors, the sensors can be mounted on both wishbones. For an IRI measurement, two laser profilometers would also be more useful.

3. Building information modelling in road asset management

3.1 Inventory

The basis for modelling the road surface is a laser scan or photogrammetric survey. In the case of a laser scan, any gaps that occur during the recording, e.g. due to parked vehicles, must be closed.

In this project, the road surface was recorded with a measuring vehicle using a modular system (Mobile Urban Mapper). This system consists of two area scan cameras with LED lighting, four still cameras, two ladybug cameras, a pavement profile scanner (PPS), a longitudinal evenness measurement system, a clearance profile scanner (CPS), and a pyrometer. The PPS scanner is used to scan the road surface over a width of 4 m. The longitudinal beams under the vehicle measure the roughness with triangulation sensors. The surface features are detected by surface cameras with LED lighting. The combination of laser scanners and cameras maps the road environment.

For further processing of the laser scan point clouds, vendor-specific programmes are required to open and process the data.

As part of the inventory, all data describing the condition of an infrastructure asset must be recorded. This includes data from the primary ordinal system (any given node-edge system, see **Figure 8**), the secondary ordinal system (areas adjacent to edges with size and structure information), and the condition (results of metrological or visual condition recording). For an existing road, this corresponds to the so-called "as-built model".

Special attention must be paid to the uniqueness and reproducibility of the primary ordering system: The nodes must be uniquely numbered so that all edges within the network also occur only once. These are designated with the adjacent nodes: Edge from node no. x to node no. y, etc.

3.2 Modelling the existing structure

To model the road body in Building Information Modelling (BIM), 3D profile bodies must be generated according to the relevant national regulations and guidelines. Attributes and properties need to be assigned to these profile bodies. For this purpose, software systems are available from various manufacturers. These

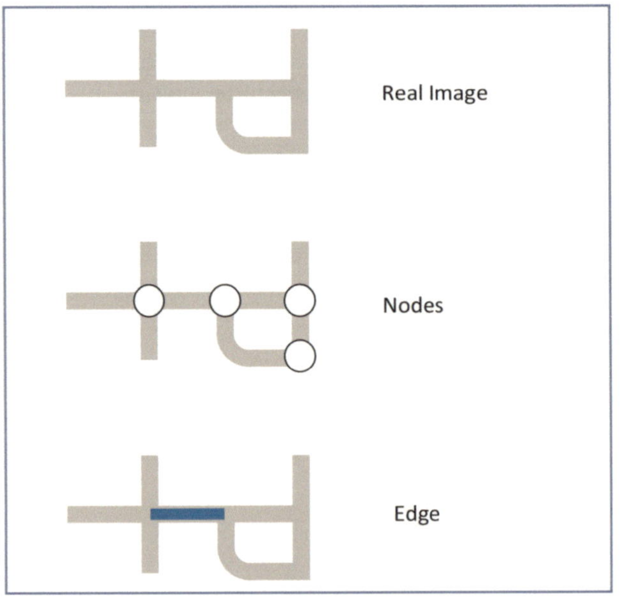

Figure 8.
Example of simple node-edge-system.

programmes are complemented by tools for visualisation, coordination, and cost and schedule planning for a comprehensive BIM application.

In this particular application, Civil 3D from Autodesk was used to model the digital twin. The available country-specific modelling packages allow the road body to be modelled quickly and in accordance with the applicable guidelines. It should be noted, however, that deviations from the normative rules and regulations that can be found in any existing structure mean that the use of these modelling packages is limited [10].

3.3 Use for operation and maintenance

The core of the road asset management application is the use case "Use for Operation and Maintenance". This standardised use case includes the transfer of data from the as-built model to corresponding maintenance management systems, the representation and, if necessary, evaluation of the condition of the structure in the model, and the updating of the model in the case of maintenance measures. These data lead to the development or updating of damage progression models, which in turn enable the planning and implementation of specific maintenance measures.

The basis for this use case is the modelled as-built model. The 3D profile body generated in this way must in turn be divided into 10 m sections to enable correlation with the condition data obtained in an earlier stage of this project. However, 10 m is not mandatory and is too short for segments on trunk and rural roads outside cities, where 100 m is more reasonable. For each of these 10 m sections here, solids were generated to which advanced properties could be assigned. In the use case, these were condition data, ordinal data, maintenance data, and acceleration data from the alternative road condition monitoring. These extended properties could be assigned to the different pavement layers in the model.

The acceleration data sets from the alternative road condition survey were divided into the same 10 m sections as the existing condition survey to allow the acceleration values to be mapped to the existing condition values. A comparison of the surface photos from the PPS scanner with the vibration values allowed the vibration values to be mapped to visually detectable road damage.

Infraworks software was used to visualise the digital twin modelled in Civil 3D. Visualisation is a convenient tool for the user to better understand the location, conditions and relationships. After importing the solids into Infraworks, it was possible to visualise condition data, as-built data, acceleration data and ordinal data.

3.4 Perspective of BIM in road asset management

It turned out that the process of creating a digital twin is user-friendly and largely uncomplicated. However, a high level of computer performance is required to be able to reproduce the surface in a detailed and realistic manner. The accuracy has an impact on the precision of the road surface and on the associated visual image of the damage. When digital models are linked to condition, maintenance and order data, the BIM working method is an enormous support for the management system. By using a digital model, data can be retrieved quickly and in greater quantity at any time. Through the volume bodies, quantity take-offs can be created automatically, which offers an advantage for resource planning and costing.

The future of BIM in road maintenance looks promising, with several potential advancements and perspectives that could significantly impact the field. Here are some key trends and perspectives to consider:

- Integration of Internet of Things (IoT) and Sensor Data: BIM can leverage data from various sensors and Internet of Things (IoT) devices embedded within road infrastructure to continuously monitor conditions. This data can be integrated with the digital twin model to provide real-time insights into road health, traffic patterns, weather impacts, and more. This integration enhances predictive maintenance capabilities, allowing authorities to address issues before they escalate.

- Advanced Analytics and Artificial Intelligence (AI): BIM combined with advanced analytics and AI will enable more accurate prediction of road degradation and deterioration patterns. AI algorithms can analyse historical data, sensor inputs, and other relevant factors to forecast maintenance needs, optimise resource allocation, and develop more efficient maintenance strategies [11, 12].

- Augmented Reality (AR) and Virtual Reality (VR) for Visualisation: AR and VR technologies can provide immersive visualisation experiences for road maintenance planning and execution. Maintenance crews can use AR glasses to overlay digital models onto the physical environment, guiding them through tasks and providing access to real-time data, reducing errors and enhancing efficiency and road availability.

- Automation and Robotics: As robotics and automation continue to advance, they can play a significant role in road maintenance tasks. Robotic vehicles and drones equipped with sensors can autonomously inspect road conditions, identifying cracks, potholes, and other issues. Automated machinery can also be used for repair and resurfacing tasks, reducing the need for manual labour. GPS data and automated guidance enable maintenance crews to easily find repair spots in large road networks.

- Lifecycle Management: BIM can support the entire lifecycle of road infrastructure, from design and construction to maintenance and final replacement. By capturing data at every stage, BIM models can provide insights into the long-term performance of materials, design choices, and maintenance strategies, leading to more informed decisions in subsequent projects.

- Digital Twins for Performance Monitoring: Digital twin technology will become more sophisticated, allowing road managers to simulate various maintenance scenarios and their potential impacts on-road performance. This proactive approach can help optimise maintenance schedules and resource allocation, leading to cost savings and improved road quality.

- Collaboration and Data Sharing: BIM promotes collaboration among various stakeholders involved in road maintenance, including road agencies, contractors, designers, and technology providers. Sharing data and insights through a centralised BIM platform can foster better communication, coordination, and knowledge exchange.

- Sustainability and Resilience: Future road maintenance strategies will likely prioritise sustainability and resilience. BIM can be used to assess the environmental impact of maintenance activities, optimise material usage, and integrate climate resilience measures into road infrastructure designs.

- Regulatory Compliance and Reporting: BIM can aid in compliance with regulations and reporting requirements related to road maintenance. Accurate and comprehensive digital records can streamline the process of documenting maintenance activities, making it easier to demonstrate compliance and respond to audits.

In conclusion, the future of BIM in road maintenance will see increased integration of advanced technologies, data analytics, and automation to create smarter, more efficient, and sustainable road networks. BIM's role will evolve from being a tool primarily used for design and construction to becoming an integral part of ongoing maintenance and infrastructure asset management.

4. Conclusion

The tried and tested method of road condition monitoring using laser scanning to measure the longitudinal and transverse profile and video recording of the road surface for subsequent signposting of road damage is labour and cost-intensive.

The multiple benefits of using vibration sensors for continuous road condition monitoring parallel the well-documented benefits of advanced sensor technologies. The ability to provide real-time insight into road conditions, enabling timely intervention, is in line with the growing trend in road management toward proactive infrastructure management using real-time data and predictive algorithms.

Still, the challenges of using vibration data for road condition monitoring are significant and varied. The following difficulties highlight the limitations and obstacles to the effective implementation of vibration monitoring in regular operations:

- Specialised Knowledge Requirements: Deriving insights from vibration data demands specialised knowledge, hindering widespread adoption.

- Limited Holistic Understanding: Vibration data alone may incompletely represent road conditions, overlooking factors like surface quality and structural integrity.

- External Influences on Data Accuracy: External factors like traffic and weather can introduce noise, questioning the reliability of vibration data.

- Lack of Context in Data: Incomplete contextual information in vibration data hampers identifying root causes, impacting maintenance prioritisation.

- Inability to Capture All Characteristics: Vibration monitoring may miss certain road conditions, potentially leaving critical issues unaddressed.

The integration of digital models with condition, maintenance and asset data offers numerous benefits that will improve decision making, efficiency and cost effectiveness. The use case demonstrated the ability to create a BIM-enabled 3D model of an existing road for road asset management, which was visualised and provided with further maintenance management attributes.

The existing examples in literature dealing with BIM application in bridge management offer a detailed exploration of the technical intricacies and strategies necessary for seamless integration. AI will play a key role in processing and analysing the vast amount of data collected from different sources. Machine learning algorithms can be trained to recognise patterns, predict road deterioration and even suggest optimised maintenance schedules based on historical data and usage patterns. By linking this information with geographic data and historical maintenance records, decision-makers can develop long-term maintenance strategies, allocate resources more efficiently, and plan for infrastructure upgrades or replacements well in advance. BIM fosters collaboration among stakeholders, promoting better communication and coordination in road maintenance efforts.

In a critical discussion, it is important to recognise that neither vibration sensors nor BIM are stand-alone solutions but can complement each other in road condition monitoring. Vibration sensors provide real-time data that can be used to complement periodic assessments. BIM provides a comprehensive data framework for informed decision making but requires consistent maintenance and engineering skills. The future potential lies in synergising these approaches to create a more holistic and efficient road maintenance strategy.

However, to deepen understanding and context, further exploration of technical challenges, integration strategies, and ethical considerations is imperative.

Integration demands addressing technical complexities such as interoperability, ensuring data standardisation, and managing scalability. Overcoming these challenges ensures a smooth amalgamation of diverse technologies. A phased, gradual implementation strategy allows for systematic adaptation and learning from pilot projects. Additionally, providing training for maintenance personnel and establishing feedback mechanisms ensures efficient utilisation and continuous improvement. Emphasis on data privacy, equity, accessibility, and transparent decision making is vital. Adhering to ethical standards builds public trust and ensures that the benefits of technology are distributed fairly across communities.

In conclusion, a holistic approach addressing technical, ethical, and practical dimensions is essential for successful integration. This comprehensive strategy maximises the potential benefits of these technologies in road maintenance, contributing to more efficient and sustainable urban infrastructure.

Author details

Berthold Best
Ohm University of Applied Sciences, Nuremberg, Germany

*Address all correspondence to: berthold.best@th-nuernberg.de

References

[1] Best B, Shtayat A, Moridpour S, Shroff A, Raol D. A review of monitoring systems of pavement condition in paved and unpaved roads. Journal of Traffic and Transportation Engineering (English Edition). 2020;7(5):629-638. DOI: 10.1016/j.jtte.2020.03.004

[2] Shtayat A, Moridpour S, Best B, Shahriar Rumi M. Using a smartphone software and a regular bicycle to monitor pavement health statues. In: Proceedings of the 2nd International Conference on Robotics Systems and Vehicle Technology 2020. New York, NY, USA: Association for Computing Machinery; 2020. DOI: 10.1145/3450292.3450312

[3] Shtayat A, Moridpour S, Best B, Rumi M. Using a smartphone software and a regular bicycle to monitor pavement health statues. In: Proceedings of the 9th International Conference on Transportation and Traffic Engineering (ICTTE); 29–31 December 2021; Naples, Italy; 2021. New York, NY, USA: Association for Computing Machinery; 2021

[4] Knuepfer M. Image Referencing in Road Condition Monitoring using Vibration Sensor Technology. Bachelor's Thesis. Ohm University of Applied Sciences Nuremberg; 2020. Nuremberg, Germany: Ohm University of Applied Sciences; 2020

[5] Angerer H. Development of Alternative Methods for Road Condition Monitoring. Master's Thesis. Ohm University of Applied Sciences Nuremberg; 2022. Nuremberg, Germany: Ohm University of Applied Sciences; 2022

[6] Best, Berthold; Angerer, Helena: Alternative methods of road condition monitoring. In: Proceedings SURF2022, the 9th Symposium on Pavement Surface Characteristics; 12–14 September 2022; Milano, Italy; 2022. Paris, France: PIARC - World Road Association; 2022

[7] Best B, Angerer H. Development of alternative methods for road condition monitoring. In: Proceedings Australasian Transport Research Forum 2022; 28–30 September 2022; Adelaide, Australia, 2022. Canberra, Australia: Australasian Transport Research Forum Incorporated; 2022

[8] Zaiko A. Correlation between the International Roughness Index (IRI) and the Vertical Acceleration of a Vehicle. Bachelor's Thesis. Ohm University of Applied Sciences Nuremberg; 2022. Nuremberg, Germany: Ohm University of Applied Sciences; 2022

[9] Abulizi K et al. Measuring and evaluating of road roughness conditions with a compact road profiler and ArcGIS. Journal of Traffic and Transportation Engineering (English Edition). 2016;**3**(3):398-411. DOI: 10.1016/j.jtte.2016.09.004

[10] Kroner K. Digital Twin for Road Asset Management. Bachelor's Thesis. Ohm University of Applied Sciences Nuremberg; 2022. Nuremberg, Germany: Ohm University of Applied Sciences; 2022

[11] Shtayat A, Moridpour S, Best B, Rumi M. An overview of pavement degradation prediction models. Journal of Advanced Transportation. 2022: Special Issue. DOI: 10.1155/2022/7783588

[12] Shtayat A, Moridpour S, Best B, Abuhassan M. Using supervised machine learning algorithms in pavement degradation monitoring. International Journal of Transportation Science and Technology. 2022;**12**(2):628-639. DOI: 10.1016/j.ijtst.2022.10.001

Chapter 2

Crumb Rubber Modified Asphalt: Fundamentals to Recent Developments

Amrita Roy, Kasilingam Rajkumar and Bharat Kapgate

Abstract

Accumulation of polymeric solid wastes is a matter of distress and concern in the modern era. Plastic wastes are easier to recycle than rubbers, especially composite products like tires are very difficult to recycle. Finding sustainable end-of-life solutions for waste rubber products is crucial at present. The use of ground rubber, popularly known as crumb rubber obtained from the shredding of waste rubber goods in the construction of road pavements is one of the best possible ways to recycle rubber. About a billion end-of-life (ELT) tires are globally discarded every year, yet the use of crumb rubber generated from these tires in road construction is still negligible. The addition of crumb rubber not only permits the consumption of a significant amount of waste rubber but also provides numerous benefits to asphalt road pavement. Crumb rubber improves the rutting resistance, fatigue cracking and durability of roads. Rubberized asphalt roads also exhibit reduced traffic noise and increased driving safety. Though there are some limitations of crumb rubber as an additive, such as phase separation and poor workability, recent technical advancements are finding ways to overcome them. Further development of this technology is crucial for the establishment of a sustainable and circular economy.

Keywords: crumb rubber, rubber modified asphalt, durability, workability, rheological properties, dynamic shear modulus

1. Introduction

Construction and maintenance of roadways in modern civilization are the two most imperative civil engineering tasks to keep the world running. Owing to the proliferating number of passenger cars, public transport and heavy-duty vehicles, road pavements are subjected to massive and quick damages. Frequent maintenance is also often uneconomical and impractical. Solutions to these issues have been explored in the past decades, mainly focusing on the modification of the conventional bituminous binder which is used with asphalt for the construction of roads. The use of polymeric modifies, especially various rubbers has been beneficial in improving the durability, flexibility and resistance to permanent deformation. The first known experimental use of natural rubber in bitumen took place almost two centuries ago, in the 1840s. The first use of synthetic rubber modifier in bitumen was reported in 1923 and the first use of waste-tire-derived rubber to modify bitumen and its consequent application in asphalt roads was reported in 1944 [1].

Nevertheless, the advancement of human civilization has started to suffocate the earth with enormous amounts of industrial, construction and consumer wastes being dumped daily. A significantly large amount of the total waste constitutes polymeric materials, either plastic or rubber. Some plastics are somewhat recyclable as they are thermoplastics. Thermoplastics can usually undergo a few cycles of melting and reshaping. Rubbers are, on the other hand, thermosets, i.e., they undergo an irreversible curing process when moulded into a specific shape. Most rubber products are cured with sulphur or other chemicals that cause the polymeric rubber molecules to form a 3-dimensional network that cannot be easily reversed [2]. Numerous moulded rubber products are used commercially and they are dumped as non-degradable solid wastes after the end of their service life. Tires are the most notable amongst the rubber products used in modern civilization. Good quality tires may last up to 5–10 years, yet, with the enormous number of passenger cars, public transport and industrial vehicles that run globally, the annual generation of waste tires is more than 1.5 billion. Scrap tires are often abandoned on the ground, becoming a sanctuary for vermin and insects leading to biohazards. Burning of scrap tires emits toxic gases into the environment. Chemicals and microplastic from tires and other rubber products also pollute the soil and water. The sustainable and hazardous solutions for scrap tires are pictorially displayed in **Figure 1**.

United Nations has recently set sustainable development goals with a focus on 'responsible consumption and production' and the 3R waste management system. Rubber and recycling industries are becoming more serious in their endeavour towards waste management. Industries are gradually developing a 4R waste management system that includes reducing, reusing, recycling and recovering. At present, 41% of end-of-life tires (ELT) still end up stockpiling and landfilling, and 42% of them are used for material recovery, 15% for energy recovery and only 2% for civil engineering [3]. Rubber recycling is possible in several techniques, such as reclamation and devulcanization, pyrolysis, downsizing or grinding. Grinding moulded rubber goods is an efficacious recycling technique that produces crumb rubber. Crumb rubber finds application in many subsidiary industries, most significantly in road pavement construction. The incorporation of crumb rubber has also remarkably

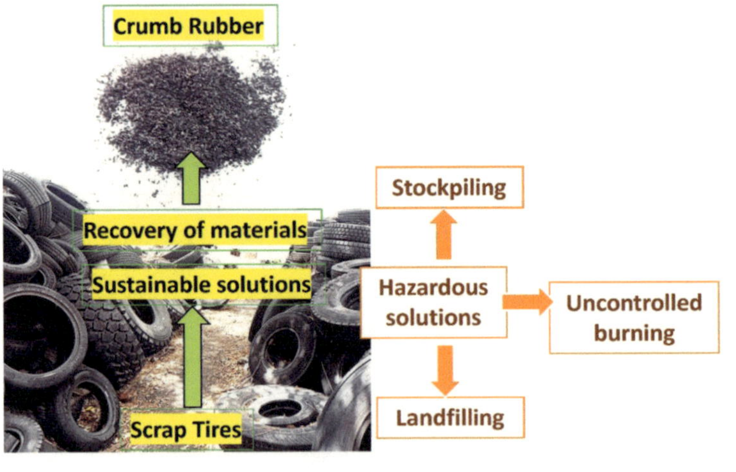

Figure 1.
The sustainable and hazardous solutions for scrap tires.

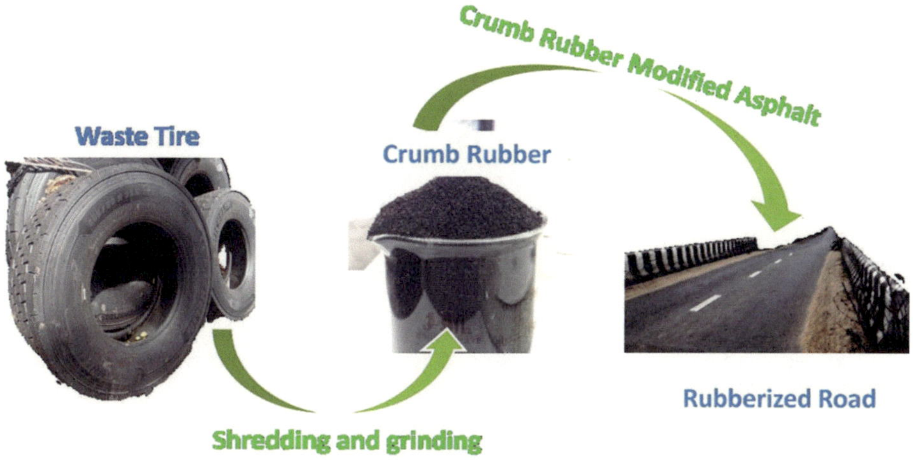

Figure 2.
Application of crumb rubber obtained from grinding of waste tires in rubberized roads.

improved the performance of asphalt in road pavements as shown in **Figure 2** [4]. Apart from being a sustainable solution for end-of-life tires (ELT), the use of tire-derived crumb rubber significantly improves the durability, temperature resistance, fatigue cracking and rutting resistance of asphalt road pavements [5]. The practise of adding crumb rubber as well as other polymer wastes such as high and low-density polyethene has been extremely advantageous for use in road pavements for improved weathering and fatigue conditions [6].

Road construction is an important component of the modern city infrastructure. Ancient roadways were built with materials like clay, sandstone or gypsum. Since the beginning of the twentieth century, roads have generally been built with asphalt which is a composite of sand and bitumen. Bitumen is a mixed hydrocarbon residue obtained from the petroleum distillation process. This highly viscous liquid works as a binder for asphalt and protects from repetitive shock or load cycles. In modern days, concrete, a mixture of Portland cement, sand and water has become a popular road construction material. In asphalt or cement concrete, two types of aggregates, typically known as coarse aggregate and fine aggregate are mixed. Fine aggregate comprises sand, clay, fly ash and fine stones that pass through a 4.75 mm sieve but retain over a 0.075 mm sieve. Coarse aggregate is retained on a 4.75 mm sieve and they contain gravel, stone and brick chips, clinker, cinder etc. The idea of incorporating rubber in bitumen for road construction emerged way back in the nineteenth century soon after natural rubber was introduced as an industrial material. The actual implementation of this idea started in the middle of the twentieth century. Ground rubber from scrap tires was added to bitumen as a modifier in 1960 in Arizona, USA. The use of crumb rubber in road materials soon became a popular trend as it enhanced the properties of the roadways. The incorporation of rubber from scrap tires is usually carried out by dissolving crumb rubber in a bitumen binder. Another way of incorporating rubber in the construction of roadways is to partially replace the fine aggregate in the asphalt or cement composite with crumb rubber. The application of ground rubber in road construction is an important step towards sustainability and is also more eco-friendly than using waste rubber for landfilling. Ground rubber provides several advantages to the performance of the roadways such as:

a. Resistance towards rutting (longitudinal depression on the road surfaces), thermal cracking and other kinds of deformations like fatigue and reflective cracking

b. Resistance towards damages caused by rain and high temperature

c. Improvement in tensile properties, resistance from stress-cracking

d. Improvement in adhesion with the aggregates, increased viscosity of the binder preventing softening of the pavement surface

e. Enhanced road grip, skid and impact resistance leading to lower possibilities of accidents and related damages

f. Increased comfort during driving and reduced traffic noise

g. Enhanced durability of the road pavements

This chapter reports a comprehensive study on rubber-modified asphalts, their preparation, properties and recent technological advancements.

2. Preparation of crumb rubber

Crumb rubber is generally produced by shredding waste rubber products. Modern grinding techniques allow even large products such as whole tires to be converted to crumb rubber. Grinding of waste rubber product is usually done in three steps as follows:

a. Primary step: Whole tires or other moulded rubber goods are downsized in a cracker mill, hammer mill or rotary shear mill. This step produces coarse rubber pieces or chips through crushing and shredding processes. The steel wires and textile reinforcing elements are also removed from the rubber.

b. Secondary step: Rubber chips are further downsized in the second step where rubber granules are formed. High-load and high-speed granulators are used in this step. Rubber pieces are often brought to a brittle and glassy state by the application of liquid nitrogen to make the cutting process easier.

c. Tertiary step: Fine ground rubber is produced in the third step by wet grinding methods.

The surface area and properties of crumb rubber depend on the temperature maintained during the grinding process. Grinding at or above ambient temperature yields crumb rubber with a larger surface area. The particle size of this crumb rubber usually ranges between 0.5 and 5 mm. Ambient grinding is also cost-effective compared to low-temperature grinding. Low-temperature or cryogenic grinding is assisted by liquid nitrogen below −80°C. The particle size of this crumb rubber is around 50 mm [7].

3. Preparation of crumb rubber-modified asphalt (CRMA)

There are two common procedures to prepare crumb rubber modified asphalt (CRMA), the dry and the wet processes. A schematic diagram of the dry and wet processes is shown in **Figure 3**.

3.1 Dry process

A portion of the fine aggregate, ideally 1–4% by weight is generally replaced with crumb rubber in this process [8]. The size of the crumb rubber particles mixed in the dry process is typically within the range of 0.4–9.5 mm. Dry mixing increases the stiffness of the mixture, making it less susceptible to stress-induced damage. Dry mixing reportedly improved the low-temperature cracking and high-temperature deformation properties of CRMA pavements [9]. The cohesion between the aggregate and asphalt is generally reduced in this dry mixing process as there is no significant interaction between the asphalt binder and crumb rubber. Crumb rubber is also said to increase the air void in the asphalt mixture in the dry process leading to damage by external factors like moisture.

However, plenty of research has been carried out to find the remedies to these issues. The use of certain reactive species to enhance the adhesion between the crumb rubber and asphalt binder has been a useful technique in the dry mixing process. An organo-silicon type nanomaterial with the commercial name of Zycosoil has been reported as an adhesion enhancer [10]. The aggregate was heated at 190°C before the addition of crumb rubber and mixed for 20 s. Varying doses of zycosoil were mixed with the asphalt binder which was preheated at 155°C. Homogeneous mixing of zycosoil was ensured by vigorous rotation of the mixture at 2500 rpm for half an hour. The softening point of the binder increased with the incorporation of the nanomaterial which improved its resistance towards high-temperature deformation. Zycosoil reportedly increased the surface free energy of adhesion between the binder and aggregate. A notable improvement in moisture resistance of the mixture was also observed as the hydrophilic aggregate surface was converted into a hydrophobic one owing to the formation of siloxane (Si-O-Si) bonds. This modification also resulted in the modulus, rutting and fatigue performance of the asphalt.

3.2 Wet process

In the wet process crumb rubber is mixed with asphalt preferably at elevated temperatures and the binder is added subsequently into the mix. Fine ground rubber,

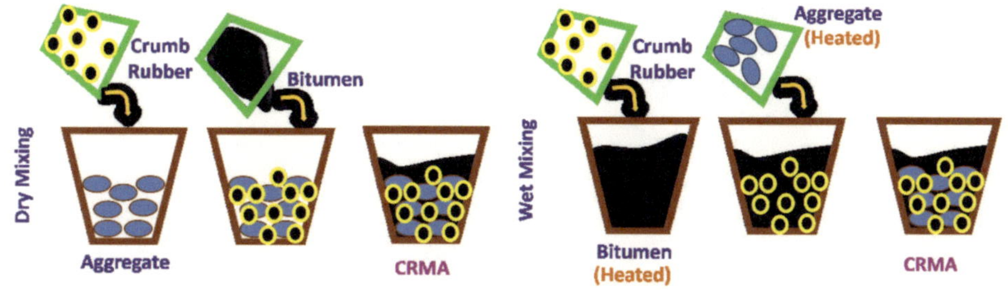

Figure 3.
Dry and wet mixing processes for CRMA.

preferably within a size range of 0.075–1.2 mm is mixed with bitumen binder. Rubber particles swell in bitumen in heated conditions inducing physical changes in the bitumen. Lighter fractions of bitumen, mostly aromatic oils, are absorbed by rubber, and the binder's viscosity increases notably. Crumb rubber modifies the binder in this process and hence this method yields significant improvement in the properties of road pavement compared to the dry process.

In a wet mixing process, the mixing sequence was studied thoroughly. The temperature of the binder was raised to 176°C when crumb rubber was added and 60 minute reaction time was given [11]. This sample was compared with a dry mixed sample. The high-temperature stability of the sample was better than the DM sample as inferred from the results of the Marshall stability test. High-temperature performance of was also superior owing to high dynamic modulus as low frequency. Conversely, the dry mix sample displayed better low-temperature performance because of its low modulus at high frequencies. The wet mix sample was also reported to display significantly higher rutting resistance in the same study.

3.3 Terminal blend asphalt rubber

Terminal blending (TB) is a modern technique that refers to a process where crumb rubber is digested by asphalt at elevated temperatures in a closed and controlled environment. Asphalt is generally heated to 200–300°C in a closed chamber where the crumb rubber is added and is subjected to high shear mixing, typically at 3000–8000 r/min speed for a sufficiently long period that enables complete and uniform blending between rubber and asphalt [12]. This procedure allows desulphurization and degradation of rubber to some extent that decreases the viscosity and enhances workability. The rheological properties of TB asphalt rubber were found to vary with the blending time as well as the temperature [13]. Longer time stimulates better dissolution of rubber and elevated temperature causes polymeric chain rupture of the rubber molecules resulting in reduced viscosity. Temperatures above 260°C cause severe chain breakage and the molecular weight of these polymer molecules is reduced to 3000–19,000 Daltons, comparable to the asphalt components. These degraded medium polymers are also known as apparent asphaltenes. The degradation of high polymers (molecular weight > 19,000 Daltons) to apparent asphaltenes aids in lowering viscosity but may deteriorate the hardness of the material over time. The stress relaxation properties of TB were found to be temperature-dependent. Shorter stress relaxation time indicates rapid dissipation of stress or high elasticity. Higher interaction temperature often results in longer relaxation time at low temperatures which is detrimental to temperature susceptibility at low-temperature ranges [14]. However, terminal blend in general provides improved workability and wide range applicability. Terminal blends consume a larger quantity of crumb rubber which has great environmental and economic benefits [15].

4. Effect of rubber on the chief properties of CRMA

The properties of rubberised bitumen depend on several variables, such as the properties and constituents of the bitumen, temperature and procedure of mixing, type of ground rubber etc. The common properties of asphalt and the effect of the addition of crumb rubber on these properties are discussed herein.

4.1 Durability

The durability of asphalt pavement indicated its resistance to ageing or age-hardening with time. The term 'durability' broadly describes a pavement's resistance to all kinds of deformations with time. Modification of bitumen with crumb rubber generally enhances the time-dependent stability as the addition of rubber increases the pavement's resilience and elasticity. Though rubberized asphalt roads may appear to be more energy-consuming due to the process of crumb rubber preparation and incorporation, rubberized roads require less frequent maintenance due to reduced wear [16]. Resistance to the damages caused by acoustic vibration is also remarkably improved in rubberized pavements [17]. Warm mix asphalt (WMA) is reported to be 50% more resistant to weathering effects. WM techniques reduced the asphalt mixing and compaction temperature by 20–40°C compared to the traditional techniques, making the process eco-friendly by reducing fuel use and emissions [18]. As the mechanical properties of the CRMA pavements are improved, a reduction in pavement thickness is possible with longer service life [19].

4.2 Workability and compaction

The workability of CRMA is the ease of mixing and compaction process of asphalt, with better workability indicating easier mixing. Workability is related to the volumetric properties as well as the pavement performance. Compaction is the process of reducing the volume of asphalt concrete during pavement construction. As the volume of the asphalt decreases, density increases and the air voids in the asphalt decrease. The incorporation of crumb rubber generally increases the compaction temperature as the air void volume increases with the addition of crumb rubber [20]. Some pieces of literature have criticised the role of crumb rubber in asphalt mixture as the absorption of the lightweight components of asphalt on the rubber surface may lead to reduced workability. However, it was also reported that if the content of crumb rubber was limited within a threshold level where the particle effect was negligible, the risk of poor workability could be avoided [21]. Controlling the viscosity of the binder at the mixing temperature is the key to achieving better workability. Road transport and highway regulations suggest that the viscosity of the modified asphalt binders should be within a limit of 300 mPa.s at 135°C. It was also observed that warm mixing techniques resulted in an improvement in the workability of CRMA. Additives like synthetic paraffin or zeolite have been reported to provide significant advantages in high-temperature performance and workability in the warm mix (WM) asphalts with low crumb rubber content [22]. The addition of the binder additives yielded low viscosity in 10–15% crumb rubber content. Another study also reported the use of commercial additives, such as Evotherm-DAT, Evotherm-3G, Sasobit etc. [23]. The Evotherm class binding additives improved workability by reducing the void content. These additives also reduced the construction temperature of CRMA by 16°C without any effect on compaction. A unique study reported the addition of natural rubber latex along with crumb rubber and organo-silane binder additive in CRMA which resulted in remarkable improvement in workability as well mechanical properties like resilience, fracture resistance, moisture resistance and reduced degrees of permanent deformation [24].

4.3 Storage stability

Crumb rubber modified asphalt (CRMA) faces the issue of phase separation of rubber from bitumen binder during the storage period. The difference between the solubility parameters of rubber and asphalt as well as the difference in density between cross-linked rubber and asphalt causes the separation or gravitational settling of rubber particles from the system. This occurrence is more frequent in the wet mixing process. It has also been observed that the larger particles have a faster rate of settling. On the other hand, more uniform compatibility and dispersion can be achieved by the use of smaller-sized rubber particles leading to superior storage stability [25]. Several attempts like the addition of crosslinking agents and modifiers have been made by researchers to increase the storage stability of CRMA. The use of styrene butadiene styrene copolymer as a modifier has been reported [26]. The formation of an internal 3D structure achieved by terminal blending at a specific time and temperature was also reported for improved storage stability [27].

4.4 Rheological properties

Rheological properties, mainly the viscosity and flow characteristics of CRMA are closely related to the workability and durability of asphalt pavement. Though very high content of crumb rubber resulting viscosity often leads to poor workability, the physical properties of asphalt binder are known to improve with the crumb rubber content. High viscosity gives rise to certain benefits such as low traffic noise, resistance to thermal damage and ageing, high rutting resistance, and longer lifetime. Rotational viscometer test is ideally used to determine the viscosity of CRMA Increased viscosity also resulted in improved penetration resistance and resistance to time-dependent deformations or creep [28]. The addition of crumb rubber to bitumen imparts a shear-thinning nature and increased elasticity. Epoxidized natural rubber (ENR) has been investigated to understand the effect of ageing on the rubberized asphalt's rheological properties [29]. The ENR-modified binders exhibited increased hardness and penetration resistance up to a certain loading of rubber. Ageing resistance was also found to increase with the addition of crumb rubber up to a certain level of rubber loading. Ductility tests of the ENR-modified asphalts displayed an initial decrease followed by an increase in ductility with increasing rubber content. An increase in ductility also indicated improved low-temperature elasticity of the asphalt pavements. The addition of trans-polyoctenamer rubber (TOR) as an additive to CRMA was reported to further improve the rheological properties and storage stability [30]. Another study also reported the benefit of using a hot recycled asphalt mixture for achieving better rheological properties. The study suggested the use of 50% recycled asphalt with 50% CRMA for optimum rheological as well as economical interest [31].

4.5 Dynamic shear modulus: Rutting resistance and fatigue cracking

Dynamic shear modulus or the complex modulus of rubber-modified asphalt binder is another critical parameter that affects the performance of CRMA. The complex modulus (G^*) is determined by the dynamic shear rheometer (DSR) test. Complex modulus (G^*) crudely indicates the bitumen binder's resistance towards deformation. The ratio of complex modulus and sin of the phase angle ($G^*/\sin\delta$) is correlated with the rutting resistance and the parameter $G^* \times \sin\delta$ is indicative of the fatigue cracking behaviour [32]. As CRMA is a viscoelastic material, it displays a non-linear relation between stress and strain, a linear region is only observed at a very low magnitude of strain. Beyond the limit of this linear region, the shear modulus reduces by almost 95% compared to its value at zero strain [33]. Crumb rubber particles, when distributed evenly in asphalt result in significant resistance towards deformation, especially to high-temperature rutting and low-temperature cracking [34]. It has been observed that the low-temperature elasticity of bitumen is enhanced in the addition of crumb rubber, but loss tangent at high temperatures may decrease resulting in lower high-temperature elasticity [35]. The incorporation of natural rubber latex as a modifier into CRMA increased the dynamic shear modulus (G*) as well as the rutting resistance at high temperatures [36]. Wax-based warm mix additives namely slack wax and polypropylene wax were used for the improvement of both rutting and fatigue cracking resistance of CRMA [37]. Similarly, Fischer-Tropsch wax-based (FT), water-based (WB) and chemical surfactant-based (CS) additives were also used to improve the rheological and dynamic properties [38]. The use of phenolic resin was reported to boost the performance of styrene-butadiene rubber (SBR) modified asphalt [39]. Phenolic resin improved the rutting resistance, and low-temperature cracking and also lowered the temperature susceptibility of CRMA. Styrene butadiene styrene (SBS) co-polymer remarkably increased the rutting resistance of CRMA when used as a binder modifier [40]. Carbon-based nanofillers like carbon nanotubes, carbon nanofibers and graphene are also being thoroughly explored to improve dynamic mechanical properties [41].

5. Factors influencing the properties of CRMA

The properties of CRMA are influenced by several factors, such as the size of the crumb rubber particles, blending time and temperature, content of rubber etc. Smaller particle size produced at high-temperature or ambient grinding results in smaller particle size which is known to have an irregular surface and larger surface area as well as better interaction with the binder. Rubber particles with cubic or regular shapes and smooth surfaces offer a smaller area for interaction with the asphalt binder. Cryogenic grinding lowers the temperature of rubber below the glass transition temperature and thus the rubber undergoes brittle fracture and smooth surfaces are produced upon grinding. On the other hand, ambient grinding generates a spongy surface which is particularly beneficial for rubber-bitumen bonding. Smaller particles also result in a dense aggregate whereas larger rubber particles produce a coarse or gap-graded aggregate. A dry mixing process with coarse rubber particles incorporating gaps between the aggregate was expected to improve the skid resistance of pavement but ultimately displayed poor resistance to low-temperature cracking [42]. A study reported a comparative analysis of crumb rubber of particle sizes ranging from 20 to 120 mesh. The content of rubber was also varied as 10, 15 and 20% by weight of the total amount of CRM-asphalt. Change in particle size changes the degree of interaction between the crumb rubber and bitumen, though the relationship is nonlinear. The softening point of CRMA displayed an increasing trend up to 60 mesh particle size with 15 and 20% rubber content and decreased after further increase in particle size. Ductility was found to be increasing with the increase in rubber content. The viscosity of CRMA decreased when the blending time was varied from 0 to 200 minutes at temperatures 180, 200, and 225°C. The ductility of CRMA was prone to increase with blending or reaction time. The study conclusively reported that crumb rubber of 80 mesh particle size displayed the best combination of properties at 15% rubber content. As the viscosity significantly reduced at elevated temperature (200°C) with blending time longer than 60 minutes, these reaction conditions were found to be optimal for fair workability [43]. Another temperature-dependent study revealed that moderately elevated temperature imparts flexibility to CRMA resulting in improved rutting resistance. The optimum temperature for the absorption of crumb rubber particles on the light components of asphalt was reported to be 190°C. Further rise in the temperature to as high as 220°C initiates intense reactions, generates new functional groups and decreases the mechanical properties and heat-stability of CRMA. Hence, it is crucial to limit the blending temperature to 190–200°C [44].

6. Challenges associated with CRMA

Regardless of the number of benefits of CRMA mentioned in this review, there are a few challenges and gaps in the development of CRMA technology. The relationship between temperature with the viscosity and workability of rubber-modified bitumen is complicated. The workability is also often dependent on manual factors. The maximum working temperature varies with different climatic conditions in different countries, e.g., 190°C in Denmark and 195°C in China. A comprehensive survey involving global participants such as the USA, India, Japan, Denmark, Australia, Canada, China etc. conducted in 2010 failed to determine a unified standard of mixing and compaction temperature (MCT) and viscosity of CRMA. Hence, there is a lacuna in the common global standard for the application of CRMA. Moreover, the reaction between rubber-modified bitumen and asphalt may continue for an indefinite period after the blending process causing untraceable changes in the properties of CRMA. There are possibilities that CRMA may not be a classically homogeneous mixture at all even after thorough blending; in most cases, the rubber particles are only encapsulated with the binders [45]. Crumb rubber contains sulphur and other organic chemicals which produce harmful smoke at elevated temperatures that are detrimental to the environment as well as the workers' health. Swelling of crumb rubber is also a common factor that increases the viscosity and reduces the workability [46]. Maintaining the storage stability of crumb rubber-modified bitumen is another practical challenge. CRMA produced by the dry process has also been reported to be susceptible to moisture damage [47].

7. Recent developments and future scope

Despite the few limitations associated with the application of CRMA, crumb rubber modification in asphalt and cement concrete appears to be indispensable for achieving sustainability in the construction sector. Recent scientific research on CRMA focuses on the limitations of its application like high viscosity, volatile content, high-temperature requirements and poor storage stability. Polyurethane was used as an additive to improve storage stability and workability and to reduce the volatile content from CRMA [42]. UiO-66 class of metal-organic complexes were reported to incredibly reduce the volatile organic contents (VOCs) resulting from the presence of unsaturated metal atoms and high surface area [43]. A blend of rubber and plastic (crumb rubber and polyethene) was reported to remarkably reduce the viscosity and improve the workability of CRMA [44]. Increasing interest is observed in the incorporation of nanomaterials like graphene and carbon nanotubes to enhance the properties of CRMA. The effect of the number of layers and size of graphene sheets on the properties of CRMA was thoroughly studied and graphene was also estimated to effectively improve high-temperature deformation and rheological behaviour [45].

The aspect of economic and environmental benefits of CRMA are in supreme focus for the future development of this technology. Refinement of the wet mixing and terminal blending methods and the use of novel additives may further reduce the viscosity and improve workability resulting in at least 20–25% fuel reserve. The emission of carbon monoxide and methane was remarkably low for CRMA pavement construction, but the emission of CO_2, NO_x, and SO_2 is still a concern that needs to be addressed. However, a quantitative study of the energy benefits of the CRMA pavements in their service life and recycling is required at this stage. The recycling and rehabilitation process of the CRMA pavements also requires further structural improvement [48].

8. Conclusions

Crumb rubber modified asphalt (CRMA) is extensively used for road construction in parts of the USA, Western Europe, Australia and South Africa. Flexible CRMA pavements are extremely durable, recyclable and require low maintenance. Crumb rubber is also a popular additive in cement-based roadways at present. It is estimated that building a kilometre-long road of 6 cm thickness with 0.5% of the total weight of bitumen modified with crumb rubber could reuse a thousand scrap tires. Replacement of bitumen with crumb rubber also significantly reduces the carbon footprint. Hence, the use of crumb rubber in road construction as well as other possible civil engineering works is probably the most sustainable solution for end-of-life tires (ELTs). Numerous long-lasting benefits of crumb rubber modification have already been established through a few decades of application and extensive research. The most advantageous feature of the rubberized roads is the crumb rubber's ability to resist fatigue cracking and plastic deformation resulting in outstanding durability and low maintenance. A 40–80% reduction in tire-pavement friction noise is also achievable by the use of CRMA. The recyclability of CRMA pavement and a significant reduction in carbon emission compared to conventional pavement construction are the two most eco-friendly features that add special value to this technology. There are disadvantages as well, such as low storage stability, rise in viscosity and phase separation of the CRMA-modified bitumen with time, difficult optimization of blending time and temperature etc., yet, the economic and environmental benefits of CRMA outweigh these process difficulties. Ample research is also going on to overcome these processing issues and the use of certain additives like paraffin wax, zeolite and co-polymers such as styrene-butadiene-styrene have been proven efficacious. Special attention is being given to the improvement of CRMA road durability in extreme and severe climate conditions as well. Further refining the CRMA technology and overcoming the few shortcomings, will be an important milestone towards sustainability. Application of CRMA and rubberized cement concrete in road construction as well as other civil engineering sectors ought to be popularised globally to establish a shared circular economy in rubber recycling and construction industries shortly.

Author details

Amrita Roy, Kasilingam Rajkumar and Bharat Kapgate*
Indian Rubber Manufacturers Research Association, Thane, Maharashtra, India

*Address all correspondence to: bk@irmra.org

References

[1] Mashaan NS, Ali AH, Karim MR, Abdelaziz M. An overview of crumb rubber modified asphalt. International Journal of Physical Sciences. 2012;7(2): 166-170. DOI: 10.5897/IJPSX11.007

[2] Adhikari B, De D, Maiti S. Reclamation and recycling of waste rubber. Progress in Polymer Science. 1 Sep 2000;**25**(7):909-948

[3] Valentini F, Pegoretti A. End-of-life options of tyres. A review. Advanced Industrial and Engineering Polymer Research. 2022;**5**(4):203-213. DOI: 10.1016/j.aiepr.2022.08.006. KeAi Communications Co

[4] Leong SY, Lee SY, Koh TY, Ang DTC. 4R of rubber waste management: Current and outlook. Journal of Material Cycles and Waste Management. 2023;**25**(1):37-51. DOI: 10.1007/s10163-022-01554-y. Springer

[5] Myhre M, Saiwari S, Dierkes W, Noordermeer J. Rubber recycling: Chemistry, processing, and applications. Rubber Chemistry and Technology. 2012;**85**(3):408-449. DOI: 10.5254/rct.12.87973

[6] Khan IM, Kabir S, Alhussain MA, Almansoor FF. Asphalt design using recycled plastic and crumb-rubber waste for sustainable pavement construction. In: Procedia Engineering. 2016. pp. 1557-1564. DOI: 10.1016/j.proeng.2016.04.196

[7] Milad A, Ahmeda AGF, Taib AM, Rahmad S, Solla M, Yusoff NIM. A review of the feasibility of using crumb rubber derived from end-of-life tire as asphalt binder modifier. Journal of Rubber Research. 2020;**23**(3):203-216. DOI: 10.1007/s42464-020-00050-y

[8] Airey GD, Rahman MM, Collop AC. Absorption of bitumen into crumb rubber using the basket drainage method. International Journal of Pavement Engineering. 2003;**4**(2):105-119. DOI: 10.1080/1029843032000158879

[9] Cao W. Study on properties of recycled tire rubber modified asphalt mixtures using dry process. Construction and Building Materials. 2007;**21**(5):1011-1015. DOI: 10.1016/j.conbuildmat.2006.02.004

[10] Arabani M, Tahami SA, Hamedi GH. Performance evaluation of dry process crumb rubber-modified asphalt mixtures with nanomaterial. Road Materials and Pavement Design. 2018;**19**(5):1241-1258. DOI: 10.1080/14680629.2017.1302356

[11] Yu H et al. Effect of mixing sequence on asphalt mixtures containing waste tire rubber and warm mix surfactants. Journal of Cleaner Production. 2020;**246**. DOI: 10.1016/j.jclepro.2019.119008

[12] Xie J, Zhang Y, Yang Y. Study on preparation method of terminal blend rubberized asphalt binder. Frontiers in Materials. 2020;7. DOI: 10.3389/fmats.2020.00279

[13] Huang W, Lin P, Tang N, Hu J, Xiao F. Effect of crumb rubber degradation on components distribution and rheological properties of terminal blend rubberized asphalt binder. Construction and Building Materials. 2017;**151**:897-906. DOI: 10.1016/j.conbuildmat.2017.03.229

[14] Li B et al. Evolution of components distribution and its effect on low temperature properties of terminal blend rubberized asphalt binder. Construction and Building Materials. 2017;**136**:598-608. DOI: 10.1016/j.conbuildmat.2017.01.118

[15] Han L, Zheng M, Wang C. Current status and development of terminal blend tyre rubber modified asphalt. Construction and Building Materials. 2016;**128**:399-409. DOI: 10.1016/j.conbuildmat.2016.10.080. Elsevier Ltd

[16] Riekstins A, Haritonovs V, Straupe V. Economic and environmental analysis of crumb rubber modified asphalt. Construction and Building Materials. 2022;**335**. DOI: 10.1016/j.conbuildmat.2022.127468

[17] Licitra G, Cerchiai M, Teti L, Ascari E, Fredianelli L. Durability and variability of the acoustical performance of rubberized road surfaces. Applied Acoustics. 2015;**94**:20-28. DOI: 10.1016/j.apacoust.2015.02.001

[18] Puccini M, Leandri P, Tasca AL, Pistonesi L, Losa M. Improving the environmental sustainability of low noise pavements: Comparative life cycle assessment of reclaimed asphalt and crumb rubber based warm mix technologies. Coatings. 2019;**9**(5). DOI: 10.3390/coatings9050343

[19] Farina A, Zanetti MC, Santagata E, Blengini GA. Life cycle assessment applied to bituminous mixtures containing recycled materials: Crumb rubber and reclaimed asphalt pavement. Resources, Conservation and Recycling. 2017;**117**:204-212. DOI: 10.1016/j.resconrec.2016.10.015

[20] Çelik ON, Atiş CD. Compactibility of hot bituminous mixtures made with crumb rubber-modified binders. Construction and Building Materials. 2008;**22**(6):1143-1147. DOI: 10.1016/j.conbuildmat.2007.02.005

[21] Yu H, Deng G, Zhang Z, Zhu M, Gong M, Oeser M. Workability of rubberized asphalt from a perspective of particle effect. Transportation Research Part D: Transport and Environment. 2021;**91**. DOI: 10.1016/j.trd.2021.102712

[22] Wang H, Li X, Xiao J, You Z, Yang X, Irfan M. High-temperature performance and workability of crumb rubber-modified warm-mix asphalt. Journal of Testing and Evaluation. 2018;**48**(4). DOI: 10.1520/JTE20170384

[23] Yu H, Leng Z, Dong Z, Tan Z, Guo F, Yan J. Workability and mechanical property characterization of asphalt rubber mixtures modified with various warm mix asphalt additives. Construction and Building Materials. 2018;**175**:392-401. DOI: 10.1016/j.conbuildmat.2018.04.218

[24] Poovaneshvaran S, Zheng LW, Hasan MRM, Yang X, Diab A. Workability, compactibility and engineering properties of rubber-modified asphalt mixtures prepared via wet process. International Journal of Pavement Research and Technology. 2021;**14**(5):560-569. DOI: 10.1007/s42947-020-1006-z

[25] Liang M, Xin X, Fan W, Sun H, Yao Y, Xing B. Viscous properties, storage stability and their relationships with microstructure of tire scrap rubber modified asphalt. Construction and Building Materials. 2015;**74**:124-131. DOI: 10.1016/j.conbuildmat.2014.10.015

[26] Cheng G, Shen B, Zhang J. A study on the performance and storage stability of crumb rubber-modified asphalts. Petroleum Science and Technology. 2011;**29**(2):192-200. DOI: 10.1080/10916460903070421

[27] Ragab M, Abdelrahman M. Enhancing the crumb rubber modified asphalt's storage stability through the control of its internal network structure. International Journal of Pavement Research and Technology. 2018;**11**(1):13-27. DOI: 10.1016/j.ijprt.2017.08.003

[28] Kim HS, Lee SJ, Amirkhanian S. Rheology investigation of crumb rubber modified asphalt binders. KSCE Journal of Civil Engineering. 2010;**14**(6):839-843. DOI: 10.1007/s12205-010-1020-9

[29] Al-Mansob RA, Ismail A, Yusoff NIM, Albrka SI, Azhari CH, Karim MR. Rheological characteristics of unaged and aged epoxidised natural rubber modified asphalt. Construction and Building Materials. 2016;**102**:190-199. DOI: 10.1016/j.conbuildmat.2015.10.133

[30] Liu H, Chen Z, Wang W, Wang H, Hao P. Investigation of the rheological modification mechanism of crumb rubber modified asphalt (CRMA) containing TOR additive. Construction and Building Materials. 2014;**67**(Part B):225-233. DOI: 10.1016/j.conbuildmat.2013.11.031

[31] Chen T, Ma T, Huang X, Guan Y, Zhang Z, Tang F. The performance of hot-recycling asphalt binder containing crumb rubber modified asphalt based on physiochemical and rheological measurements. Construction and Building Materials. 2019;**226**:83-93. DOI: 10.1016/j.conbuildmat.2019.07.253

[32] Palit SK, Reddy KS, Pandey BB. Laboratory evaluation of crumb rubber modified asphalt mixes. Journal of Materials in Civil Engineering. Feb 2004;**16**(1):45-53

[33] Petersen JC, Robertson RE, Branthaver JF, Harnsberger PM, Duvall JJ, Kim SS, et al. Binder characterization and evaluation: Volume 1. Rep. No. SHRP-A-367, Strategic Highway Research Program. Washington, DC: National Research Council; May 1994

[34] Wang L, Xing Y, Chang C. Microscopic and dynamic rheological characteristics of crumb rubber modified asphalt. Journal Wuhan University of Technology, Materials Science Edition. 2010;**25**(6):1022-1026. DOI: 10.1007/s11595-010-0142-8

[35] Navarro FJ, Partal P, Martínez-Boza F, Gallegos C. Influence of crumb rubber concentration on the rheological behavior of a crumb rubber modified bitumen. Energy and Fuels. 2005;**19**(5):1984-1990. DOI: 10.1021/ef049699a

[36] Poovaneshvaran S, Mohd Hasan MR, Putra Jaya R. Impacts of recycled crumb rubber powder and natural rubber latex on the modified asphalt rheological behaviour, bonding, and resistance to shear. Construction and Building Materials. 2020;**234**. DOI: 10.1016/j.conbuildmat.2019.117357

[37] Ameri M, Afshin A, Ebrahimzadeh Shiraz M, Yazdipanah F. Effect of wax-based warm mix additives on fatigue and rutting performance of crumb rubber modified asphalt. Construction and Building Materials. 2020;**262**. DOI: 10.1016/j.conbuildmat.2020.120882

[38] Kataware AV, Singh D. Dynamic mechanical analysis of crumb rubber modified asphalt binder containing warm mix additives. International Journal of Pavement Engineering. 2019;**20**(9):1044-1054. DOI: 10.1080/10298436.2017.1380806

[39] Ming LY, Feng CP, Siddig EAA. Effect of phenolic resin on the performance of the styrene-butadiene rubber modified asphalt. Construction and Building Materials. 2018;**181**:465-473. DOI: 10.1016/j.conbuildmat.2018.06.076

[40] Kumar SA, Veeraragavan A. Dynamic mechanical characterization of asphalt concrete mixes with modified asphalt binders. Materials Science and Engineering A. 2011;**528**(21):6445-6454. DOI: 10.1016/j.msea.2011.05.008

[41] Li Y et al. Anti-rutting performance evaluation of modified asphalt binders: A review. Journal of Traffic and Transportation Engineering (English Edition). 2021;**8**(3):339-355. DOI: 10.1016/j.jtte.2021.02.002. Chang'an University

[42] Hassan NA, Airey GD, Jaya RP, Mashros N, Aziz MA. A review of crumb rubber modification in dry mixed rubberised asphalt mixtures. Jurnal Teknologi. 2014;**70**(4):127-134. DOI: 10.11113/jt.v70.3501

[43] Cao WD, Liu ST, Cui XZ, Yu XQ. Effect of crumb rubber particle size and content on properties of crumb rubber modified (CRM) asphalt. Applied Mechanics and Materials. 15 Oct 2011;**99**:955-959

[44] Yang X, Shen A, Guo Y, Lyu Z. Effect of process parameters on the high temperature performance and reaction mechanism of CRMA. Petroleum Science and Technology. 2018;**36**(19):1537-1543. DOI: 10.1080/10916466.2018.1482315

[45] Mohammadi I, Khabbaz H. Challenges associated with optimisation of blending, mixing and compaction temperature for asphalt mixture modified with crumb rubber modifier (CRM). Applied Mechanics and Materials. 15 Feb 2013;**256**:1837-1844

[46] Zheng W, Wang H, Chen Y, Ji J, You Z, Zhang Y. A review on compatibility between crumb rubber and asphalt binder. Construction and Building Materials. 2021;**297**. DOI: 10.1016/j.conbuildmat.2021.123820. Elsevier Ltd

[47] Ghabchi R, Arshadi A, Zaman M, March F. Technical challenges of utilizing ground tire rubber in asphalt pavements in the United States. Materials. 2021;**14**(16). DOI: 10.3390/ma14164482

[48] Wang T, Xiao F, Zhu X, Huang B, Wang J, Amirkhanian S. Energy consumption and environmental impact of rubberized asphalt pavement. Journal of Cleaner Production. 2018;**180**:139-158. DOI: 10.1016/j.jclepro.2018.01.086. Elsevier Ltd

Chapter 3

Strengthening Substandard Road Materials with Nanoemulsion-Based Stabilisation: An Experimental Study

Eche S. Okem, Nathan N. Chilukwa and Mohamed M.H. Mostafa

Abstract

This study considered using nanotechnology-based products to stabilise and enhance the properties of a substandard quality road construction material. Two stabilisers, herein referenced as Nano-A and Nano-B, were utilised to stabilise a G6 gravel material with a California bearing ratio (CBR) of 25% and a Plasticity Index (PI) of 4%. X-ray diffraction (XRD) tests showed that the material contained 58% Quartz with 36.5% clay minerals in the fraction passing the 0.075 mm sieve. The impact of the stabilisers on the material was evaluated through CBR, unconfined compressive strength (UCS), and indirect tensile strength (ITS) tests. The results revealed that both stabilisers significantly improved the bearing strength of the material by as much as 53 and 92% for Nano-A and Nano-B, respectively. Significant improvements were also noted in UCS and ITS results. The study also revealed that both rapid and 28-day curing conditions result in similar strength effects in the samples. Nano-B also showed great potential for improving the material's hydrophobic properties as measured by the resistance to water damage. The research findings indicate an enhanced load-bearing capacity of the stabilised material, ensuring long-lasting pavement infrastructure. These enhancements demonstrate the effectiveness of Nano-A and Nano-B in improving the properties of substandard road construction materials. Tests on more substandard materials are recommended.

Keywords: nanoemulsion, substandard materials, stabilisation, hydrophobic, road materials

1. Introduction

It is necessary to use construction materials of sufficient quality and strength when building roads. However, as such materials are not readily available, alternative strategies are considered to improve available substandard materials to ensure their utilisation in road construction through various stabilisation techniques. Road material stabilisation improves substandard material properties through increased mechanical strength and load-bearing capacity. The selection of an appropriate stabiliser depends on several factors, including the material's properties, compatibility, and desired improvements [1].

In South Africa (SA), road materials are classified as G1 to G10, referencing the quality of the road materials based on various properties particularly, the strength [2]. G1 to G3 are composed of high-quality crushed stone road materials, while G4 to G10 refer to natural granular materials ranging from gravel to silts and clays [2, 3]. G1 to G3 are costly and not readily available for road construction compared to G4 to G10, which often requires stabilisation to enhance their performance when utilised as road construction material.

Traditional stabilisers such as cement, lime, bitumen, and fly ash are some of the most commonly used products for road material stabilisation [4, 5]. However, these products often present challenges, including developing brittle materials susceptible to cracking and compatibility with certain material types, e.g., materials containing mica and clay minerals [6–10]. The limitations associated with these traditional stabilisers in road construction objectify the need to seek or consider alternative products suitable for the current demand.

Over the years, several alternative (non-traditional) products have been appraised with regard to their capacity to enhance substandard road materials [11, 12], including nanotechnology-based stabilisers. These stabilisers exploit nanomaterials' unique properties to control the materials' properties at the nano-scale [13, 14]. Advances in nano-science and nanotechnologies have made it possible to use this technology for pavement engineering [3, 7, 15], with initial trials concentrated on asphaltic materials to improve binder performance and durability [16, 17]. However, recent studies have shown that nanotechnology-based stabilisers can also be applied to granular materials for the construction of pavement structural layers. This includes the utilisation of bitumen emulsions modified with nanotechnology to effectively stabilise substandard road materials [3, 6, 7]. The modification of bitumen emulsions with nanotechnology improves tensile strength and provides resistance to deformation in road materials compared to un-modified bitumen emulsions [18].

Numerous studies [3, 6, 7, 18–24] on bitumen emulsions modified with nanosilanes indicated that these products have a unique advantage of drawing on the properties of nanotechnology to improve road construction material properties by forming siloxane bonds, one of the strongest bonds in nature [25]. These products are hydrophobic and thus impart this property to stabilised road materials such that they become water-resistant [18, 25]. Water resistance is an invaluable property, particularly for road materials found in tropical regions, which are susceptible to secondary weathering effects, usually through exposure to moisture. In this case, the bitumen emulsion acts as a carrier fluid, aiding the transportation and penetration of the nanomaterial into the intergranular network of the road material. Additional benefits of utilising bitumen emulsions modified with organofunctional nanosilanes in road material stabilisation include providing adhesion between particles and encapsulating fine aggregates.

Similar to the road material classification presented in TRH14 [2], encompassing categories from G1 to G10, cement-stabilised materials (C1 to C4), and bituminous stabilised materials (BT1 to BT3), Jordaan et al. [6] propose a classification system for road materials incorporating nanomaterial stabilisation, termed nano-modified emulsion (NME1 to NME4 and NM-EG5). In this classification, the "1" suffix denotes the highest quality road material, with subsequent suffixes indicating diminishing material quality. These classifications hold significant importance as they facilitate a meaningful comparison between the road material categorisation outlined in TRH14 [2] and that of road materials stabilised with nanomaterials.

As mentioned above, a number of commercial solutions have been developed and launched as workable nanotechnology-based alternatives to traditional bitumen emulsion in road material stabilisation. Therefore, a thorough evaluation of these solutions is required before considering their incorporation into standard processes for building roads.

A report elsewhere [26] indicates an abundance of substandard road materials in several locations in the province of KwaZulu-Natal that are unsuitable for road construction. KwaZulu-Natal is a provincial location in South Africa. It has a varied climate ranging from coastal humidity to inland aridity and different types of soils, including sandy soils or clay-rich soils, making it appropriate for studying the effectiveness of various road stabilisation methods in these complex environments. A research gap exists to investigate the potential of improving the performance of these substandard road materials to ensure their utilisation in building quality roadways through stabilisation with nanoemulsions as an alternative to traditional stabilisers.

This research considered two commercially available products for the stabilisation of substandard road materials. The stabilisers were modified bitumen emulsions incorporating nanotechnology products, herein referred to as Nano-A and Nano-B.

Nano-A is a non-traditional water-based modified bitumen emulsion with nano-polymer, and according to the product's datasheet, the stabilising polymer of Nano-A has elastomers that require sufficient compaction to bind them to the soil matrix, enabling it to gain strength once cured. The curing procedure in Nano-A is achieved through moisture loss in the mix, a concept similar to reverse osmosis – moisture escapes from inside the mix into the atmosphere. Still, it prevents water from penetrating, forming a water-resistant layer as curing occurs. Nano-A requires ultraviolet light to enhance its strength property.

Nano-B is a bitumen emulsion modified with organofunctional nanosilane, enabling the fusing of soil granules with the nanosilane. As the emulsion breaks and water is expelled from Nano-B stabilised road material, a hydrophobic effect will be created in the material. There is limited information regarding the effectiveness and test procedure in stabilising road materials with Nano-A, unlike Nano-B, which has gained some attention in this regard. Furthermore, the recommendation and procedure for testing Nano-B are contained in the Draft document TRH24 [27] and will be adopted for this study.

The research aims to compare the performance of these two modified bitumen emulsions for stabilising substandard road materials for the construction of road structural layers. The study also assessed the materials' resistance to moisture damage as well as the effect of rapid and 28-day aged curing methods.

The importance of this study is to ensure the use of *in situ* substandard road material in road construction through stabilisation with nanoemulsions, which will be instrumental in minimising the need to haul good quality road materials to the construction site.

2. Materials and test methods

Three materials were used in this study. The first is a G6 natural gravel road material obtained from Stockville quarry in Pinetown, KwaZulu-Natal, Eastern

South Africa (SA). The others are two nanotechnology-modified bitumen emulsions, Nano-A and Nano-B. The choice of these stabilisers stems from their availability in the research region as nanotechnology products for road material stabilisation. Furthermore, both Nano-A and Nano-B are modified bitumen emulsions with nanotechnology and are claimed to have the capability to provide waterproving protection for road material. It is appropriate to subject both products to similar test conditions to evaluate and provide a fair comparison of their effectiveness in stabilising road material. Therefore, as discussed below, the test recommendation and procedure for nano-modified emuslion (NME) were employed for this investigation.

Classification tests were conducted on the road materials in accordance with relevant SA road material test methods, as shown in **Table 1**. The mineral composition of the road material and its compatibility with Nano-A and Nano-B are evaluated in Sections 3.1 and 3.2. Therefore, X-ray diffraction (XRD) scans were conducted on a whole sample of the material and on a sample of a fraction passing the 0.075 mm sieve per the literature review [20, 21].

The two stabilisers used are proprietary products bound by copyright restrictions, so no chemical analysis tests were conducted on them. However, their physical properties are provided in **Table 2**.

The gravel material was stabilised with 0.7 to 1.5% of Nano-A and Nano-B. The trends in mechanical strength properties were evaluated and compared. The test procedures espoused in Draft TRH24 [21, 27] for NME were adopted for material

Name of test	Aims	Test methods
Sieve analysis	To classify the construction material	TMH1, Method A1(b) [28]
Atterberg limits		TMH1, Method A2, A3 [28]
MDD/OMC	Strength determination	TMH1, Method A7 [28]
CBR		TMH1, Method A8 [28]
Indirect tensile strength		TMH1, Method A8 [28]
Unconfined compressive strength		TMH1, Method A14 [28]

Table 1.
Test methods for the investigation.

Property	Nano-A	Nano-B
Colour/appearance	Dark grey	Thick, milky brown liquid
Flammability	None	None
Odour	Slightly	Slightly
Solubility in water	Yes	Yes
Toxicity	None	None
Boiling point	Same as water	Same as water
pH	8–9	10–11

Table 2.
Properties of the research stabilisers.

specimen preparation, which rely primarily on the guidelines in TMH1 and TG2 [28, 29]. Tests conducted included CBR, ITS and UCS, as presented in **Table 1**. Six specimens were prepared for each test, with three specimens tested for relevant properties immediately after curing. In contrast, the other tests were conducted after 4 hours of wet conditioning (soaking in water).

Air-drying specimens in the laboratory for 28 days can be time-consuming and lead to construction delays. However, any curing period can be used for testing, with 7, 28, and 90 days being the most common [30]. This study compared the two curing methods, rapid curing and air-drying for 28 days, by testing the UCS and ITS of specimens of Nano-A. Only Nano-A specimens were tested for both curing conditions, while Nano-B specimens were tested after rapid curing. This is attributed to testing limitations. The various strength tests performed on the specimens are shown in **Figure 1**.

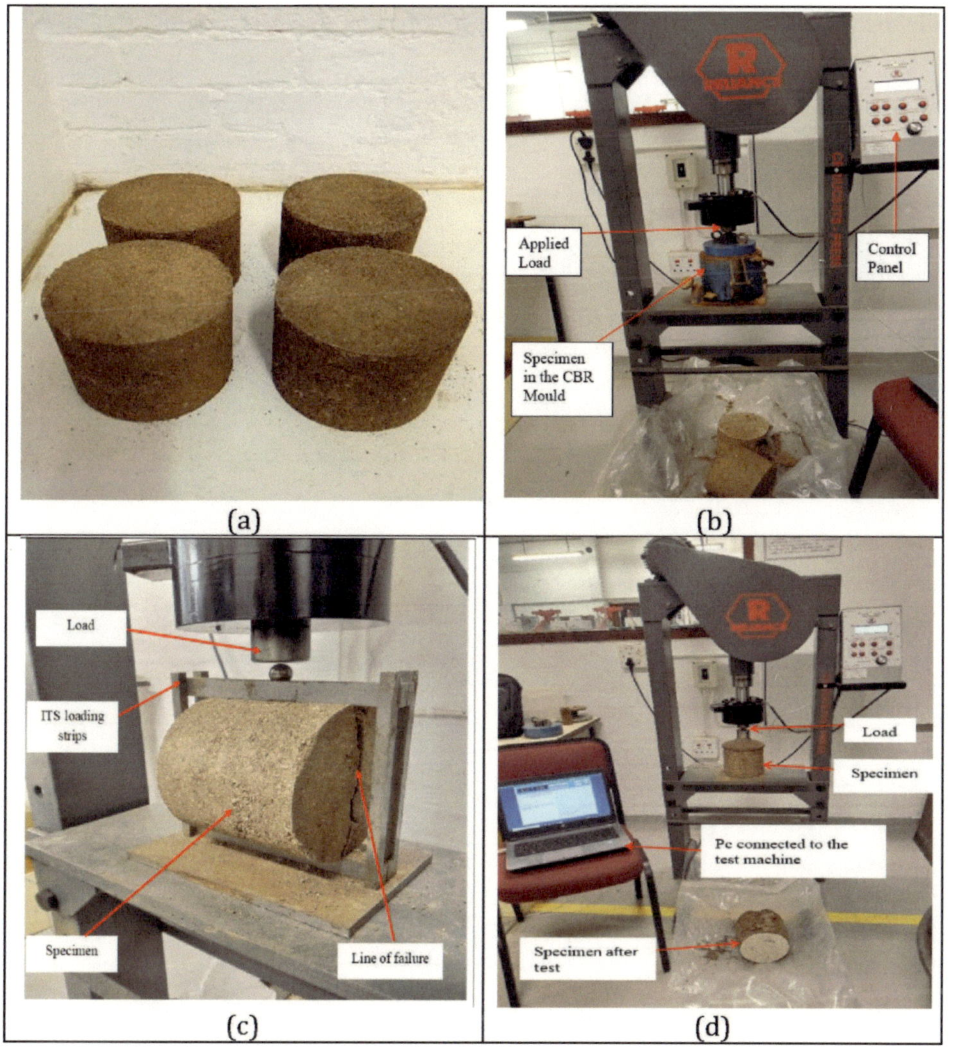

Figure 1.
(a) Specimens, (b) CBR test arrangements, (c) ITS test arrangements, and (d) UCS test arrangements.

3. Results and discussion

3.1 Material classification

The particle size distribution curve of the road material is shown in **Figure 2**. The maximum aggregate size is 37.5 mm, and its grading coefficient G_c is 22.42. The material has a Grading Modulus (GM) of 2.3. **Table 3** provides the results of additional classification tests conducted on the road material. Based on the results and referencing SA standards [2, 31], it is determined that the road material is a G6, which is relatively substandard natural gravel for pavement construction.

3.2 XRD test

The mineralogy report for the road material from the XRD test is presented in **Table 4**. The test on the sample indicates that it is predominated by quartz (58%), while the fraction passing the 0.075 mm sieve contains 36.5% clay mineral (23.9%

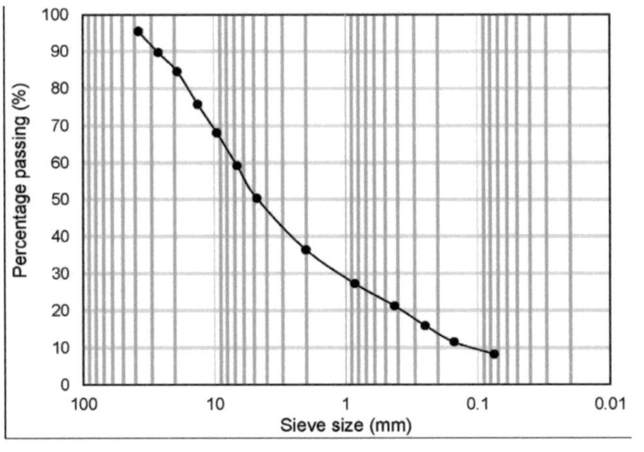

Figure 2.
Particle size distribution curve for the road material.

Liquid limit (%)	30
Plastic limit (%)	26
Plasticity index (PI)	4
Linear shrinkage	—
Coarse (%)	49.7
Sand (%)	42
Clay and silt (%)	8.14
Optimum moisture content (%)	5.8
Maximum dry density (kg/m^3)	2148
CBR at 93 Mod AASHTO (%)	25

Table 3.
Characteristic properties of the road material.

Mineral Groups	Total Sample	0.075 mm Fraction
	Primary mineral – Not subject to weathering	
Quartz (%)	58	20.4
Total	58	20.4
	Primary mineral – Subject to weathering	
Plagioclase (%)	20.1	17.8
Microline (%)	20.6	16.9
Diopside (%)	0.1	0
Haematite (%)	0	0.7
Total	40.8	35.4
	Mica	
Muscovite (%)	1	7.7
Total	1	7.7
	Clay Minerals	
Kaolinite (%)	0.2	23.9
Smectite (%)	12.6	12.6
Total	12.8	36.5

Table 4.
Mineralogy report for the road material.

Kaolinite and 12.6% smectite) and 7.7% mica. The quartz, a silica tetrahedron, is necessary to form siloxane bonds compatible with the stabilisers [32]. Clay and mica (especially the muscovite sub-group) are among the undesirable minerals in construction materials as they tend to absorb and retain water in compacted layers [2]. The combined quantity of clay minerals and mica (44.2%) is critical in determining the quantities of stabiliser necessary to mitigate their impact.

3.3 Mixing procedure for the stabilisation

The appropriate application rate for the stabiliser was selected following the recommendation of Jordaan and Steyn [21]. The authors recommend the application rates based on the CBR of the unstabilised material, i.e., 32% (at 95% of Mod AASHTO density), the quantity of the material passing the 0.075 mm sieve (8.14%), and the quantity of the secondary minerals identified through the XRD tests, 44.2%. It is shown in **Figure 3**, based on these results, that the recommended stabiliser by mass lies in Zone 2. Based on the preceding details, the application range for both products, Nano-A and Nano-B, was 0.7 to 1.5%, with Nano-A applied at 0.7, 1, and 1.5%, while Nano-B was applied at 0.7, 1, 1.2, and 1.5%. ITS and UCS tests were performed at the specific application ranges for Nano-A and Nano-B. The CBR test was carried out at the optimum content determined from the ITS and UCS tests for Nano-A and Nano-B.

3.4 CBR test

The results of CBR tests on specimens stabilised with 1 and 1.2% of Nano-A and Nano-B, respectively, are presented in **Figure 4**. Nano-B had the most significant

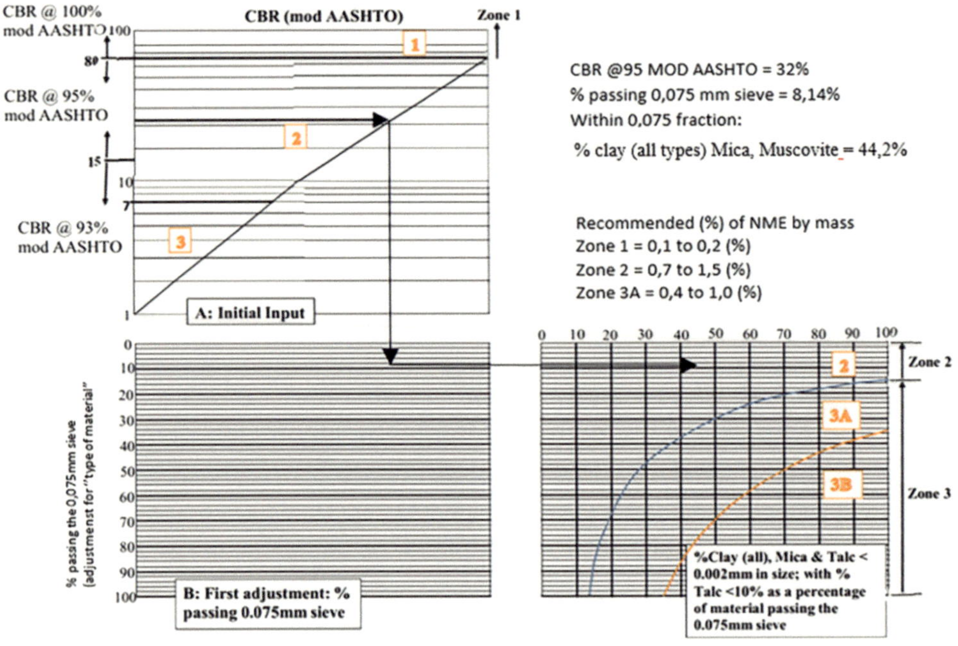

Figure 3.
Selection of an appropriate nano-modified emulsion technology for naturally available road material. Based on Jordaan and Steyn [21].

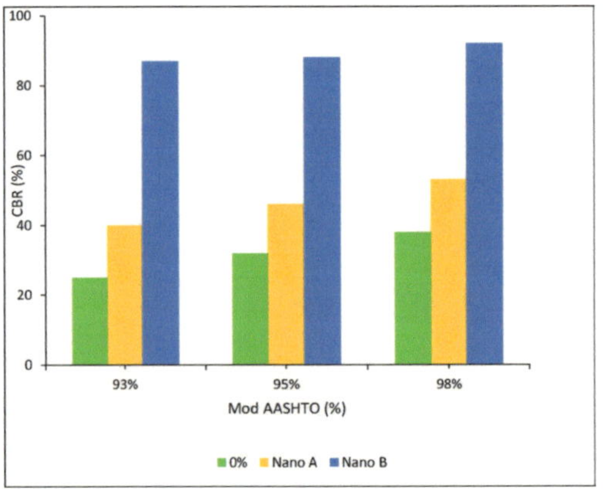

Figure 4.
The California bearing ratio for the stabilised and unstabilised specimens at 93, 95 and 98% Mod AASHTO.

impact on the soaked CBR value due to its hydrophobic effect, which created a firm interlocking of the particles in the specimens. This was not the case for Nano-A stabilised material. The soaked CBR test is a crucial attribute for evaluating the bearing capacity of soil material on site when it is in contact with a water source (surface or underground).

The CBR of unstabilised material at 93 and 95% Mod AASHTO met the minimum CBR requirement of 25% for a G6 material. However, the CBR of Nano-A stabilised material at 98% Mod AASHTO was below the minimum CBR requirement of 80% for a base course layer. On the other hand, Nano-B stabilised material exceeded the CBR requirement even at 93% Mod AASHTO. It is important to note that CBR tests are irrelevant for material selection for modified bitumen emulsions with nanotechnology. Still, they are included to accommodate traditional methods of material selection [21].

3.5 ITS test

The results of ITS tests for specimens stabilised with Nano-A and Nano-B at 0.7 to 1.5% are shown in **Figure 5**. Both Nano-A and Nano-B improved the performance of ITS specimens compared to the unstabilised specimens. The ITS_{dry} results for unstabilised specimens cured rapidly are 2.4% lower than those cured for 28 days. Specimens stabilised with 0.7% Nano-A cured rapidly have ITS_{dry} results that are 2.8% lower than those cured for 28 days in the air. The maximum ITS_{dry} results of 152 and 168 kPa were achieved for 1% Nano-A specimens cured rapidly and for 28 days in air, respectively. ITS_{dry} results of 149 and 161 kPa were achieved for 1.5% Nano-A specimens cured rapidly and for 28 days in air, respectively. The optimum content for Nano-A is 1%, as the strength attained at this content is greater than at 1.5%. Specimens stabilised with 0.7% Nano-B cured rapidly have ITS_{dry} results of 148 kPa, while the peak ITS_{dry} result of 152 kPa was achieved at 1.5%. The ITS_{dry} results for Nano-A specimens increase with increasing content, while the strength attained at 1 and 1.2% with Nano-B specimens is slightly lower than the strength at 0.7%.

Another crucial aspect of this study is determining the specimens' capacity to resist moisture damage during the 4 hours of soaking, which was evaluated through the retained tensile strength (RTS), as expressed in Eq. (1) [21, 28]. RTS measures the effectiveness of the tensile strength of the materials, expressed as a percentage of the wet and dry ITS specimens:

$$RTS = \frac{ITS_{wet}}{ITS_{dry}} (\%) \tag{1}$$

During the 4 hours of soaking the specimens in water, it was observed that the unstabilised specimens dissolved after about 30 minutes. Samples stabilised with Nano-A at 1 and 1.5% remained stable throughout the 4 hours of soaking in water and still were not tested for ITS_{wet} due to water absorption in the specimens, while

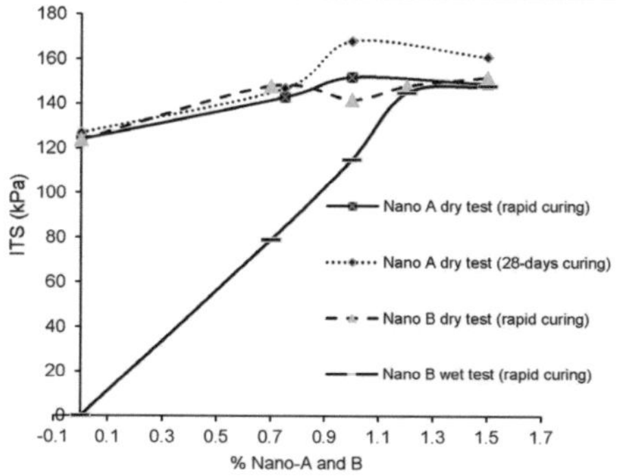

Figure 5.
ITS for the stabilised and unstabilised specimens.

specimens stabilised at 0.7% dissolved after about 2 hours of soaking. The ITS_{wet} of samples stabilised with Nano-B increases as the content of the stabiliser increases from 0.7 to 1.5%. At 1.2 and 1.5% of Nano-B, the ITS_{wet} design requirement in draft TRH24 [28] (>120 kPa) for NME4 (nano-modified emulsion) stabilised material was satisfied. A retained tensile strength (RTS) of 53.4, 81.0, 97.9%, and 97.4 were obtained at 0.7, 1, 1.2, and 1.5% of Nano-B. Therefore, the optimum content for Nano-B is 1.2% since no significant increments in strength at 1.5% were recorded. The ITS results for Nano-A and Nano-B showcased how they affect the strength properties of the specimens in the dry and wet states.

3.6 UCS test

As shown in **Figure 6**, UCS results for specimens stabilised with Nano-A and Nano-B and unstabilised specimens that were cured rapidly or for 28 days in the air are compared. The results for the UCS_{dry} (dry test) show a difference of 3.8% between the two curing methods, with the air-cured specimens recording 1.83 MPa and the rapid-cured specimen 1.76 MPa. Similar results were observed with the specimen stabilised with 1% Nano-A. A UCS_{dry} of 2.29 MPa was obtained for the air-cured Nano-A stabilised specimens, while 2.17 MPa was obtained for the rapid-cured, resulting in a difference of 5.3%. No significant improvement was recorded as Nano-A was increased from 1 to 1.5% in the specimens. The observed difference can be attributed to moisture variation in the specimens due to the curing conditions.

The results also show that the UCS_{dry} (dry test) for Nano-B performs better than Nano-A specimens, although a little decline in the UCS_{dry} was observed at 1 and 1.2% with Nano-B. The observed variation in strength between Nano-A and Nano-B can be related to their unique chemical composition.

The resistance to moisture damage of the specimens was also evaluated through the Retained Compressive Strength (RCS), expressed in Eq. (2) [21, 28]. RCS is a measure of the effectiveness of the compressive strength of the materials, expressed as a percentage of the wet and dry UCS specimens:

$$RCS = \frac{UCS_{wet}}{UCS_{dry}}(\%) \tag{2}$$

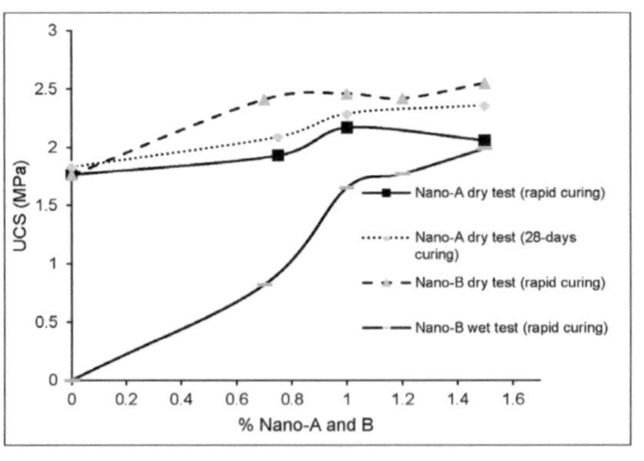

Figure 6.
UCS of specimens stabilised with Nano-A and Nano-B.

It was observed that the unstabilised UCS specimen dissolved within 30 minutes of being soaked in water for 4 hours. At 0.7%, Nano-A specimens dissolved within 2 hours, while those of 1 and 1.5% were stable throughout the 4 hours of soaking and still not be tested due to excessive moisture in the specimens.

RCS of 34.0, 67.0, 73.0, and 78.0% were obtained at 0.7, 1, 1.2, and 1.5% for specimens stabilised with Nano-B. UCS_{wet} for Nano-B specimens exceeds the UCS_{wet} design requirement of 0.75 MPa for NME4 stabilised material. In contrast, the minimum requirements of 65% for RCS were exceeded at 1 to 1.5% for Nano-B specimens, which indicates the specimen's resistance to deterioration in wet conditions.

Overall, Nano-A provides no significant protection to specimens during the 4 hours of soaking in water, while Nano-B creates a hydrophobic effect in the specimen's matrix during the soaking. However, UCS_{dry} results for Nano-A and Nano-B exceed the requirement by TRH4 [2] for a cemented natural gravel (C4) base course layer equivalent to natural stabilised material with nano-modified emulsions of category 4 (NME4). In terms of structural capacity, the NME4 stabilised layer can withstand a traffic loading between 1.0 MESA to 3.0 MESA [28]. These results indicate that Nano-B has the potential to improve the performance of stabilised road material in wet conditions, while Nano-A provides no significant improvement in this regard. The variations in the performance of specimens stabilised with Nano-A and Nano-B in wet conditions are closely related to their chemical compositions.

3.7 Conclusion and recommendations

This study compared the effectiveness of two nanotechnology-based stabilisers, Nano-A and Nano-B, in improving the strength of substandard G6 gravel material for use in a pavement structure. The results demonstrated that both Nano-A and Nano-B, at concentrations ranging from 0.7 to 1.5%, significantly improved the UCS_{dry} and ITS_{dry} of the specimens.

Notably, Nano-A performed well in enhancing the ITS dry specimens, while Nano-B exhibited the highest effect on the UCS dry specimens. Furthermore, the minimum RTS and RCS requirements were met with Nano-B concentrations at 1, 1.2, and 1.5%. The superior performance of Nano-B in wet conditions can be attributed to the presence of nanosilane in its composition, resulting in a hydrophobic effect on the specimen. The elastomers of Nano-A did not provide significant protection against water absorption in the specimens and, hence, failed to satisfy the moisture resistance capacity as mentioned in its product's datasheet.

Interestingly, increasing the concentration of Nano-A from 1 to 1.5% did not significantly improve ITS_{dry} and UCS_{dry} results. On the other hand, Nano-B concentrations of 1.2 and 1.5% created similar effects on ITS_{dry} and ITS_{wet} results, but a slight improvement in UCS_{dry} and UCS_{wet} was observed as the Nano-B content increased from 1.2 to 1.5%. Thus, the optimum content for Nano-A and Nano-B was determined to be 1 and 1.2%, respectively.

The study's insights into the curing process open new avenues for optimising construction schedules. The similarity in strength outcomes between rapid and 28-day curing conditions suggests that significant time savings could be achieved without compromising material quality.

The study findings indicated that the enhancement in UCS_{dry} with Nano-A and Nano-B exceeded the base course requirements for cemented-stabilised natural gravel material of category 4 (C4). Based on the CBR results at 95% Mod AASHTO, the Nano-A stabilised sample is suitable for the subbase, while the Nano-B stabilised sample is adequate for the base course at 98% Mod AASHTO. Applying these nanotechnology-enhanced stabilisers to substandard paving materials would facilitate the use of *in situ* materials for building quality roads.

Considering the investigation and analysis performed in this study, several recommendations are proposed for future research:

1. Further research is necessary to investigate the long-term interactions of the stabilisers with the road material.

2. This research was performed on only one road material type (G6); therefore, it cannot be generalised to all substandard materials; hence, further studies on other substandard road materials are recommended.

3. Variations in temperature, humidity, and wind and their effects on air-cured specimens should be evaluated to understand better and establish a correlation between air-cured and rapid-cured samples.

4. The effect of these stabilisers on the performance of materials with higher plasticity should be investigated to explore their potential for broader applications.

5. Lastly, the recyclability of the material stabilised with these stabilisers should be evaluated within the context of the circular economy and their compatibility with existing recycling equipment.

Addressing these recommendations would contribute to a deeper understanding of the stabilisers' capabilities and allow for more informed decision-making in implementing nanotechnology-based solutions for pavement materials. Ultimately, these advancements would promote sustainable infrastructure development and contribute to the use of *in situ* materials, leading to the construction of durable and high-quality roads.

Acknowledgements

The authors would like to express their sincere gratitude to the manufacturers of the stabilisers for granting permission and support to conduct this investigation. Additionally, the authors extend profound appreciation to the University of KwaZulu-Natal, where the study was conducted, for providing the necessary resources and facilities that made this research possible. Special thanks are also due to the management of Giba Gorge Business Park and Soilform for their invaluable cooperation in supplying the gravel material used in the study. Their contributions have been instrumental in the successful completion of this research project.

Conflict of interest

The authors declare no conflict of interest in this study.

Author details

Eche S. Okem*, Nathan N. Chilukwa and Mohamed M.H. Mostafa
Sustainable Transportation Research group (STRg), University of KwaZulu-Natal, Durban, South Africa

*Address all correspondence to: samokem333@gmail.com

References

[1] Mwaipungu RR, Allopi D. The appropriate material specifications and manual are key for effective gravel road design, construction, and maintenance practice. In: Proceedings of the First National Conference on Intermodal Transport: Problems, Practices, and Policies. Hampton, Virginia, USA: Eastern Seaboard Intermodal Transportation Applications Center, Hampton University. 2012. pp. 176-188

[2] Technical Recommendation for Highways (TRH) 14. Guidelines for Road Construction Materials. Pretoria: CRSA; 1986

[3] Jordaan GJ, Wynand J. Constructability of road pavement layers using new-age (nano) modified emulsion (NME) stabilisation of naturally available granular materials in roads varying from highways to local access roads. 2021. DOI: 10.20944/preprints202110.0181.v2

[4] Jones D, Rahim A, Saadeh S, Harvey JT. Guidelines for the Stabilization of Subgrade Soils in California. California, United States: University of California Pavement Research Center, UC Davis and UC Berkeley; 2011

[5] Heyns M, Mostafa Hassan M, Abejide SO. Sustainable waste alternative as cement replacement in pavement stabilisation. Journal of Construction Project Management and Innovation. 2018;**8**(1):1767-1778. DOI: 10.10520/EJC-10a49a808d

[6] Jordaan GJ, Kilian A, Du Plessis L, Murphy M. The development of cost-effective pavement design approaches using mineralogy tests with new nano-technology modifications of materials. In: Proceedings of the 36th Southern African Transport Conference. Pretoria, South Africa: Southern African Transport Conference; 10-13 July 2017. Available from: http://www.satc.org.za/assets/1a_jordaan.pdf [Accessed: April 15, 2020]

[7] Rust FC, Smit MA, Akhalwaya I, Jordaan GJ, Du Plessis L. Evaluation of two nano-silane-modified emulsion stabilised pavements using accelerated pavement testing. International Journal of Pavement Engineering. 16 Apr 2022;**23**(5):1339-1352. DOI: 10.1080/10298436.2020.1799210

[8] Kidgell M, JVDM Steyn W, Jordaan G. Effect of nano-modified emulsions (NME) (Nano-Silanes) stabilisers on the properties of dolomite. In: Proceedings of the Southern African Transport Conference. Pretoria, South Africa: Southern African Transport Conference; 2019

[9] Sebesta S. Use of microcracking to reduce shrinkage cracking in cement-treated bases. Transportation Research Record. 2005;**1936**(1):2-11

[10] Li X. Shrinkage cracking of soils and cementitiously-stabilized soils: Mechanisms and modeling [dissertation]. Washington State University; 2014

[11] Moustafa A, Bazaraa A, Nour El Din A. Soil stabilisation by polymeric materials. Macromolecular Materials and Engineering. 1981;**97**(1):1-2

[12] Al-Khanbashi A, El-Gamal M. Modification of sandy soil using water-borne polymer. Journal of Applied Polymer Science. 2003;**88**(10):2484-2491

[13] Rogers B, Adams J, Pennathur S. Nano-technology: Understanding Small Systems. Boca Raton, Florida: CRC Press; 2014

[14] Ganesh VK. Nano-technology in civil engineering. European Scientific Journal. 2012;**8**(27):96-109

[15] Bahari M, Nikookar M, Arabi M, Haghi AK, Khodabandeh H. Stabilization of silt by nano-clay. In: Proceedings of the 7th National Congress on Civil Engineering. Zahedan, Iran: University of Sistan and Baluchestan; 7-8 May 2013. Available from: https://www.researchgate.net/publication/236676576 [Accessed: June 18, 2022]

[16] Partl MN, Gubler R, Hugener M. Nano-science and-technology for asphalt pavements. Special Publication - Royal Society of Chemistry. 2004;**292**:343-356

[17] Van De Ven MF, Molenaar AA, Besamusca J, Noordergraaf J. Nano-technology for binders of asphalt mixtures. In: Proceedings of the Fourth Eurasphalt and Eurobitume Congress, 2008 May. Copenhagen, Denmark: Euopean Asphalt Pavement Association (EAPA); 2008

[18] Jordaan GJ, Steyn WJ. Fundamental principles ensuring successful implementation of new-age (nano) modified emulsions (NME) for the stabilisation of naturally available materials in pavement engineering. Applied Sciences. 2021;**11**(4):1745. DOI: 10.3390/app11041745

[19] Akhalwaya I, Rust FC. Laboratory evaluation of road construction materials enhanced with Nano-Modified Emulsions (NMEs). In: Proceedings of the 37th Annual Southern African Transport Conference (SATC 2018) and International Conference on Transport Infrastructure (ICTI). Pretoria, South Africa: Jukwaa Media; 9-12 July 2018. Available from: https://www.researchgate.net/publication/326480143_laboratory_evaluation_construction_materials_enhanced_with_nano-modified_emulsions_nmes [Accessed: April 29, 2020]

[20] Jordaan GJ, Kilian A, Muthivelli N, Dlamini D. Practical Application of Nano-technology in Roads in Southern Africa. In: Proceedings of the 8th Africa Transportation Technology (T2) Conference; 2017. Lusaka, Zambia. Available from: http://www.soilform.co.za/pdf/T2%20conference%20in%20Livinston%20-%20Final%20Paper%2016-05-2017.pdf. [Accessed: December 21, 2022]

[21] Jordaan GJ, Steyn WJ. Nano-technology incorporation into road pavement design based on scientific principles of materials chemistry and engineering physics using new-age (Nano) modified emulsion (NME) stabilisation/enhancement of granular materials. Applied Sciences. 2021;**11**(18):8525. DOI: 10.3390/app11188525

[22] Jordaan GJ, Steyn WJv. Engineering properties of new-age (nano) modified emulsion (NME) stabilised naturally available granular road pavement materials explained using basic chemistry. Applied Sciences. 2021;**11**:9699. DOI: 10.3390/app11209699

[23] Kidgell MM, Steyn WJ, Jordaan G. Effect of nano-modified emulsions (NME) (nano-silanes) stabilizers on the properties of dolomite. In: Proceedings of the Southern Africa Transportation Conference (SATC 2019). Pretoria, South Africa: Southern African Transport Conference; 2019. Available from: https://www.researchgate.net/publication/334443389 [Accessed: July 22, 2020]

[24] Rust FC, Akhalwaya I, Jordaan GJ, du Plessis L. Evaluation of a nano-silane modified emulsion stabilised base and subbase under HVS traffic. In: Proceedings of the 12th Conference on Asphalt Pavements for Southern Africa

(CAPSA'19), Sun City, South Africa. 2019. Available from: http://hdl.handle.net/10204/11220 [Accessed: August 5, 2020]

[25] Wheeler G. Alkoxysilanes and the consolidation of stone. Los Angeles, California, USA: The Getty Conservation Institute; 2005. Available from: https://www.getty.edu/conservation/publications_resources/pdf_publications/pdf/alkoxysilanes_vl.pdf [Accessed: December 7, 2020]

[26] Bennett HE, Ducasse K, Payne GA, Sewlal S. Innovative initiatives in road design and construction in the province of KwaZulu-Natal, South Africa. In: Proceedings of the 21st Annual South African Transport Conference 'Towards Building Capacity and Accelerating Delivery', 2002 July 15-18. South Africa. Pretoria: SATC; 2002. ISBN: 0-620-28855-8

[27] Technical Recommendation for Highways Draft (TRH) 24. Upgrading of Unpaved Roads. Pretoria, South Africa: The South African National Roads Agency SOC Limited; Nov 2022

[28] TMH1. Standard methods of testing road construction materials. Pretoria: CSIR; 1986

[29] Technical Guideline (TG2). Bitumen Stabilised Materials: A guideline for design and construction of bitumen emulsion and foamed bitumen stabilised materials. 2nd ed. South Africa: Asphalt Academy; 2009. ISBN 978-0-7988-5582-2

[30] ASTM D5102-09. Standard Test Methods for Unconfined Compressive Strength of Compacted Soil-Lime Mixtures. West Conshohocken, PA: ASTM International; 2009

[31] Committee of Land Transport Officials (COLTO). Guidelines for Road Construction Materials; South African Institution of Civil Engineers (SAICE). South Africa: Halfway House; 1985

[32] Jordaan GJ, Steyn W. Testing of granular/soil characteristics for the optimisation of pavement designs using reactive stabilising agents including "new-age" nano-technologies. In: Proceedings of the 12th Conference of Asphalt Pavements for Southern Africa (CAPSA 2019). South Africa: Sun City; 2019. pp. 13-16

Chapter 4

A Comprehensive Scientific Theory for Hot Mix Asphalt Design Aiming Accuracy and Simplicity for Asphalt Pavement Construction and Mix Quality Check

Jung Huh

Abstract

This chapter deals with the first attempt to derive a scientific, revolutionary, hot mix asphalt design method. The full rutting equation was derived as a function of the damage wheel pass number (N_d) in the previous paper by Huh. Since N_d is hard to measure, the single stroke breakdown energy (W_{1d}), is disclosed to replace N_d. By using this W_{1d}, new performance equations are proposed as a function of the W_{1d}. The effectiveness of the new linear performance equation derived is experimentally validated. These performance equations in the form of master equations could constitute a scientific, rigorous, hot mix asphalt design theory. Since the formulation of master equations requires much effort and time, a simple design variable representing master equations is proposed as a design variable. This is an important discovery in this paper. Comparison of the suggested variable (W_{1d}) among various asphalt mixes can predict both rutting and fatigue cracking simultaneously. It is hoped that this noble HMA design method can be a promising alternative for currently available empirical methods like *Marshall, Gyratory Compactor, Indirect Tensile, BMD,* etc. to obtain a better design value.

Keywords: full rutting equation, damage wheel pass number, modified fatigue cracking equation, one-stroke damage energy, master performance equations of asphalt mixes

1. Introduction

All asphalt mix design methods available so far have been empirical for the past 90 years. This paper first time deals to derive the scientific hot mix design method. The gap between empirical and scientific methods is found that the formal methods rely on experiences and observations, while the latter method relies on scientific findings reaching performance equations and experimental validation of those equations. This means that this novel hot mix design method depends on performance equations of asphalt mixes, not experiences yielding empiricism.

1. Introduction – review of the previous finding.

2. Development of New Asphalt Mix Performance Equations.

3. Validation of Mix Performance Equation Containing W_{1d}

4. Master Performance Equations for Rigorous Design Eqs.

5. A Novel HMA Design Theory for the Asphalt Pavement Construction and the Asphalt Mix Quality Check

6. Conclusion

The compacted asphalt concrete mix at the rutting temperature (40–80°C) is a typical multiphase material comprised of solid phase (aggregate and filler), viscous melt phase (asphalt binder), and air phase (air void). Thus, the melted asphalt mix becomes a strong multiphase material that usually forms the capillary pressure due to presence of the interphase tension between solid (aggregate and filler) and viscous melt medium (asphalt binder), and various particle curvatures. The capillary pressure force in the multiphase materials creates the network solid structure by combining every binder-coated particle together. Thus, the solid structure of compacted asphalt mixes requires repeated loadings (or wheel pass numbers) to breakdown the solid structure formed at the breakdown (0r the damage) point. The nature of the solid network structure formation of the compacted asphalt mix itself at the rutting temperature is the same as the one at the fatigue cracking temperature except only temperature difference.

The existence of the capillary pressure force in a multiphase material has been studied before. Gillespie and Settineri [1] computed an adhesion force comprising a liquid medium bridge connecting two identical spherical particles by using the capillary pressure force. Mehrotra and Sastry [2] also derived a mathematical expression of the adhesion force for a pendular liquid bridge medium connecting two unequal particles. These papers and others claim the existence of the capillary pressure force in multiphase materials.

The full rut performance equation in the term of a damage wheel pass number (N_d) has been derived in the view of breakdown (or damage) of the solid structure. A detailed description is given in the earlier paper by Huh [3] where Eq. (1) and Eq. (2) below are derived to explain the full rut depth quantitatively. Eq. (1) tells that the full rut depth at a fixed wheel pass number (h_N) is only a function of the damage wheel pass number (N_d) which is a characteristic property of a given mix. In addition, Eq. (1) is claimed to be independent of wheel pass speed, temperature, and asphalt binder viscosity (Huh [3]). According to Eq. (1), the full rut depth at a fixed N (h_N) can be determined if N_d is known.

$$h_N = Log\left(\frac{m}{N_d{}^B}\right) \cdot N^{\beta} + C \cdot \left[Log\left(\frac{m}{N_d{}^B}\right)\right]^D \cdot N^{\delta} \tag{1}$$

Here, h_N is the full rut depth at a fixed N, N is the number of wheel passes, N_d is the wheel pass number reaching a damage point (or a damage wheel pass number), and B, C, D, b, and d, are material constants.

All usual rutting variables affecting the full rut depth at a fixed N (h_N) are hidden in the N_d variable as shown in Eq. (2).

$$N_d = F(\text{measuring device, wheel weight, wheel speed, temperature, asphalt mix property}). \tag{2}$$

In Eq. (2), the asphalt mix property in the parenthesis is characterized by physical variables such as aggregate (kind, size, and gradation), asphalt binder (viscosity, elasticity, degree of oxidation, and adhesion), filler (kind and size), and air void (percentage) of a given asphalt mix. Thus, the estimation of N_d by Eq. (2) is difficult due to the many variables involved. Kaloush and Witczak [4] who spent much time and effort to evaluate N_d experimentally reported that its evaluation was time-consuming, and repeatability was not to be assured. In predicting the full rut depth by Eq. (1), direct estimation of N_d is preferred, instead of using Eq. (2).

In Eq. (2), if the measuring device, wheel speed, wheel weight, and temperature are fixed to certain values, then, N_d measured solely represents the asphalt mix property for a given mix as shown in Eq. (3).

$$N_d = F(\text{asphalt mix property}). \tag{3}$$

2. Development of new asphalt mix performance equations

The N_d in Eq. (1) under the condition of Eq. (3) is nothing but the structural damage wheel pass number characterized by the property of a given asphalt mix. It is proposed that the number of wheel passes reaching the damage point (N_d; or a damage wheel pass number) in the solid structure of rut deformation is identical to the number of loadings required in reaching the fatigue cracking failure (N_f) under the form of Eq. (1). Note that both involve breakdown of the solid structure, and the only difference is temperature.

Numerous empirical fatigue cracking models are available in the fatigue cracking study for asphalt concrete mixes up to now. Among them, two major types of fatigue cracking models have received much attention. One relates fatigue cracking to the strain and the modulus of the asphalt concrete mixes, as shown in the work of Finn et al. [4]. The other relates the fatigue cracking failure to the dissipated energy consumed up to the structural damage point. In this study, the latter model is chosen for modification because the general solid structure study tells that energy dissipation is usually involved in its failure.

Dijk [5] studied the accumulated dissipated energy per unit volume up to the structural failure point (W_{fat}) with respect to the repetition number of applied small load in its structural failure (N_{fat}), where the bending beam specimens of asphalt concrete mixes in the solid-state are tested at the room temperature. He reported that, for a given asphalt mix, both W_{fat} and N_{fat} are related to each other in the form of Eq. (4) shown below, and this equation is governed independently with respect to the test method (2- or 3-point bending beam test), temperature change (10–40°C), frequency change (10–50 Hz), mode of deformation control, or magnitude of the load applied. That is,

$$W_{fat} = A \cdot N_{fat}{}^{Z} \tag{4}$$

where A and z are physical constants, N_{fat} is the repetition number of the applied load at the fatigue failure point, and W_{fat} is the total dissipated energy accumulated by a small incremental energy in reaching the structural failure point. Dijk and Visser [6] studied Eq. (4) further by fitting more data obtained from the bending beam failure test (at −10 to 10°C), where specimens were made of all kinds of asphalt concrete mixes. In their tests, temperature was not fixed and possible material changes were not considered. Here, in this study, the energy-related failure model of Dijk [3], Eq. (4), is chosen as the fatigue cracking model. But it is modified for measurement to be easy, time to be saved, and effort to be less. It is assumed that the ratio of accumulated dissipated energy at failure (W_{fat}) among various asphalt mixes is proportional to the ratio of the single stroke breakdown energy (W_{1d}) among the same mixes. That is,

$$\mathbf{W_{fat}}\left(=\sum_{i=1}^{N_d}\mathbf{W_i}\right)\propto W_{1d} \tag{5}$$

Here, W_i is the incremental, dissipated energy at each repetition. By the assumption, N_{fat} corresponds to N_d, and W_{fat} matches to W_{1d}. Now, Eq. (4) can be changed into the form of Eq. (6).

$$\begin{gathered} N_d = \left(\frac{1}{A}\right)^{1/z} \cdot W_{fat}{}^{1/z} = P \cdot W_{1d}{}^{R} \\ \mathrm{Log}\,(N_d) = \mathrm{Log}\,(P) + R \cdot \mathrm{Log}\,(W_{1d}) \end{gathered} \tag{6}$$

where N_d is the wheel pass number reaching the structural damage point, W_{1d} is the energy required to breakdown of a pavement concrete mix specimen by a single stroke, and P (= $(1/A)^{1/z}$) and R (= 1/Z) are material constants. Eq. (6) is the new fatigue cracking performance equation based on the single stroke breakdown energy (W_{1d}), which is claimed, first time here. Eq. (6) could eliminate the uncertainty involved in incremental testing by Eq. (4). Eq. (6) indicates that the accumulated incremental fatigue cracking equation (Eq. (4)) is changed into the single stroke fatigue cracking equation (Eq. (6)).

Now, one wants to insert the new fatigue cracking damage expression, Eq. (6), into the rut damage wheel pass number (N_d) in the full rutting equation, Eq. (1), to change the N_d variable hard to measure into W_{1d} simple to estimate. Then, one obtains the full rutting performance equation, Eq. (1), in terms of the single stroke breakdown energy (W_{1d}). The resultant equation is demonstrated in Eq. (7). Note that N is fixed here.

$$H_N = [S\text{–}T\,\mathrm{Log}(W_{1d})]\,N^{b} + C\,[S\text{–}T\,\mathrm{Log}(W_{1d})]^{D}\,N^{d} \tag{7}$$

where h_N is the full rut depth (h) at a fixed N, W_{1d} is the single stroke breakdown energy, and S, T, C, D, b, d are constants. Eq. (7) shows that the number of wheel passes at the damage point (N_d) in Eq. (1), which is difficult to measure and time-consuming in measurement, has been changed into W_{1d} in Eq. (7), which is relatively easy to measure. Note that a fatigue cracking variable (W_{1d}) is incorporated into the full rut depth equation. This is the natural consequence because both rutting and fatigue cracking involve breakdown of the solid structure that is built on the same capillary pressure at different temperatures. Thus, the full rutting equation, Eq. (7), works if and only if the fatigue cracking equation, Eq. (6), works well. This implies that the single stroke damage energy (W_{1d}) evaluated for the fatigue cracking performance equation at the fatigue cracking temperature range (or 10–20°C) can be used to evaluate the full rutting performance equation (Eq. (1)) for the same mix at the rutting temperature range (or 50–70°C) as well. They are interrelated based on Eq. (7).

The full rut depth expression in Eq. (7) works at a fixed N that is chosen arbitrarily among the entire N. For the rut depth comparison, the wheel pass number at the damage point (N_d) is a desirable choice for N, because it represents a unique property of a given asphalt mix in the full rutting curve. The linear ((a)) and the full ((b)) rutting deformations, and N at N_d are demonstrated in **Figure 1**.

Thus, Eq. (7) can be rewritten in terms of the structural damage point by using Eq. (6) at $N=N_d$, as

$$H_{Nd} = [S\text{–}T\ \mathrm{Log}(W_{1d})]\ N_d^{\ b} + C\ [S\text{–}T\ \mathrm{Log}(W_{1d})]^D\ N_d^{\ d} \quad (8)$$

Eq. (8) can be rewritten as Eq. (9) by inserting Eq. (6), $N_d = P{\cdot}W_{1d}^{\ R}$, into Eq. (8),

$$H_{Nd} = [E - F\ \mathrm{Log}(W_{1d})]\ W_{1d}^{\ j} + C\ [E - F\ \mathrm{Log}(W_{1d})]^D\ W_{1d}^{\ k} \quad (9)$$

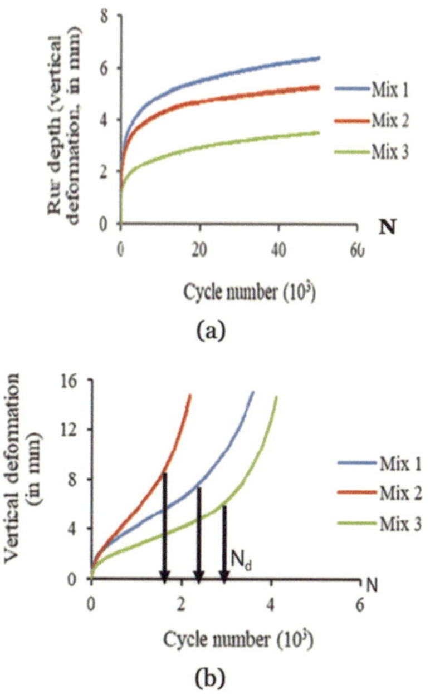

Figure 1.
(a) Linear rut depth, (b) Full rut depth, and $N = N_d$ *for different asphalt mixes [7].*

Since the rut depth at the damage point does not deviate much from the consolidated deformation curve (h_1) mentioned in the reference [1], the approximate form of Eq. (9) can be written by modifying the consolidated deformation term [1] as

$$h_{Nd} \approx [E - F\,\mathrm{Log}(W_{1d})]^{O}\, W_{1d}{}^{P}. \qquad (10)$$

If a small N (=N_s) immediately after the post compaction region is chosen, the rut depth solely corresponds to the consolidated deformation (h_1) [1] with the negligible dissipated deformation (h_2) [1]. In this case, further simplification is made to yield only the linear consolidated deformation [1]. That is,

$$h_{Ns}/N_s{}^{b} \approx h_{Ns} = S\text{–}T\,\mathrm{Log}(W_{1d}). \qquad (11)$$

Eq. (11) is the linear rut depth equation for different asphalt mixes at the fixed small number of wheel passes immediately after the post compaction region. Note that even in this simplest rutting equation, the equation holds the property relating to solid fatigue cracking variable (W_{1d}), different from the traditional linear rutting equation ($h = \alpha N^{\beta}$).

At crack temperatures (0°C-(−40°C)), some molecules in straight (or modified) asphalt binders of asphalt mixes will undergo phase changes such as local crystallization or glass transition. Once any phase change in asphalt mixes occurs at low temperatures, their physical properties become different from those of the original ones with no phase change. Thus, special attention is required when using the fatigue cracking equation (Eq. (6)) at low temperatures.

Eq. (6) can be used for either cracks or low-temperature cracks with W_{1d} that have changed due to phase change in the asphalt binders. Thus, at each low temperature of (−15°C), (−30°C), or (−45°C), W_{1d} for all asphalt mixes should be measured to find the low-temperature performance because measuring temperatures are all different, regardless of phase change. Of course, rutting is out of concern at these temperatures, and only the fatigue cracking equation (Eq. (6)) must be considered.

Note that Eq. (11) is a simple linear rutting equation to prove its effectiveness. For the validation study, a laboratory wheel tracking device is used to measure the rut depth at N = 2100 (a relatively small number of wheel tracking), performed at 60°C, while the single stroke damage energy (W_{1d}) is measured at 20°C which belongs to the fatigue cracking temperature. Note that both rutting and fatigue cracking properties are simultaneously involved even in this simple rutting equation.

3. Validation of a rut performance equation containing W_{1d}

3.1 Data measured by the laboratory wheel tracking test

The aggregate size and gradation used are shown in **Table 1**. The 10,450 g (95.0 wt.%) of the aggregates and 550 g (5.0 wt.%) of the modified asphalt binder are used to prepare a single mix specimen (11 kg) (100 wt.%) for the laboratory wheel tracking experiment. The aggregate weight percentage includes 2.0 wt.% of fillers (the limestone powder). Aggregates were heated to 190°C in a temperature-controlled oven for 2 hours to eliminate all moisture.

The modified asphalt binder is prepared by adding 3, 4, 5, and 6% of the polymer additive to 97, 96, 95, and 94% of the straight asphalt binder (AC-20), and all are mixed in a high shear, small mixer at 170°C for 1 hour. By mixing 13 mm porous and

19 mm modified asphalt mixes		13 mm porous asphalt mixes	
Aggregate size (mm)	**Weight %**	**Aggregate size (mm)**	**Weight %**
19 ~ 13.2	25.00%	—	—
13.2 ~ 9.5	41.08%	13.2 ~ 9.5	48.89%
9.5 ~ 4.75	17.88%	9.5 ~ 4.75	31.08%
4.75 ~ 2.36	2.84%	4.75 ~ 2.36	4.41%
2.36 ~ 0.6	3.68%	2.36 ~ 0.60	5.64%
0.6 ~ 0.3	1.60%	0.60 ~ 0.30	2.26%
0.3 ~ 0.15	1.77%	0.30 ~ 0.15	2.34%
0.15 ~ 0.07	1.02%	0.15 ~ 0.075	1.11%
Filler	5.13%	Filler	4.26%
Sum	100%	Sum	100%
Binder Content	5.8%	Binder Content	4.2%

Table 1.
Aggregate size and distribution.

19 mm regular aggregates (10,450 g) with four different modified binders (550 g) in a high shear mixer at 170°C, four different 19 mm modified mixes and four different 13 mm porous asphalt mixes are prepared.

These are poured into the wheel tracking mold (30 × 30 × 5 cm) at 160°C and are compacted 55 times by the compaction roller with a diameter of 46 cm, compaction pressure of 5 kg/cm^2, and tracking speed of 23 cm/s. The compact height is consistent such that the air voids of the compacted mixes are in the range of 4 to 6%. Before performing wheel tracking experiments, the prepared specimens in the tracking mold are cured for more than 18 hours at room temperature and then put in the environmental chamber at 60°C for 6 hours to achieve a homogeneous temperature distribution in the mixed specimen.

After 6 hours of specimen exposure at 60°C, the wheel tracking test begins inside the environmental chamber at 60°C. In the test, the width of the steel tracking wheel is 5 cm, its diameter is 20 cm, and the wheel pressure is 70 ± 1 kg. The wheel tracking distance is 23 ± 1 cm, and the speed is 42 ± 1 passes per minute. At 21, 105, 210, 315, 630, 945, and 1260 cycles of wheel tracking, the deformation of the specimens is measured and recorded. Dynamic stability is obtained by computing the slope of the deformation curve in the tracking time range of 45–60 minutes. The test results are shown in **Tables 2** and **3**.

3.2 Data measured by Marshall and indirect tensile testing

Marshall specimens for both kinds of asphalt concrete mixes (13 mm porous asphalt concrete mixes and the 19 mm modified ones) are prepared in Section 3.1 above. Each asphalt concrete mix specimen was designed to be 1100 grams. The 13 mm porous asphalt mix specimen required 4.2% modified asphalt binders, while the 19 mm modified mix required 5.8% modified asphalt binders. The heated aggregates and a modified binder to make one specimen were poured into the Marshall mixing bowl and mixed for 3 minutes, and then the mix was back into the oven at 170° C. After 1 hour in the oven, it was poured into the Marshall mold and compacted 50 times on both sides by the Marshall compactor to make a Marshall specimen. The

13 mm porous asphalt mixes							
Measurement			**Wheel tracking depth (mm)**				**Remarks**
			1 spec	**2 spec**	**3 spec**	**4 spec**	
Number of wheel passes		25	0.52	0.42	0.44	0.45	
		105	0.90	0.70	0.72	0.72	
		210	1.06	0.84	0.86	0.87	
		315	1.16	0.92	0.95	0.95	
		630	1.33	1.07	1.10	1.14	
		945	1.43	1.15	1.19	1.25	
		1260	1.50	1.21	1.25	1.34	
Regression results by $h = \alpha N^{\beta}$							
α			0.2997	0.2300	0.2372	0.2168	
β			0.2293	0.2360	0.2360	0.2563	
r^2			0.9930	0.9953	0.9963	0.9992	
test item		Unit	Analysis results				Specification
			1	2	3	4	
Wheel tracking evaluation	Dynamic Stability	Passes/mm	9000	10,500	10,500	7000	KSF-2374: 2000
	Deformation Rate	mm/minute	0.0047	0.0040	0.0040	0.0060	

Table 2.
Wheel tracking test results for 13 mm porous mix specimens.

19 mm modified asphalt mixes							
Measurement			**Wheel tracking depth (mm)**				**Remarks**
			1 spec	**2 spec**	**3 spec**	**4 spec**	
Number of wheel passes		25	0.40	0.41	0.47	0.25	
		105	0.61	0.64	0.73	0.47	
		210	0.69	0.73	0.85	0.59	
		315	0.76	0.79	0.93	0.65	
		630	0.85	0.87	1.04	0.76	
		945	0.90	0.93	1.11	0.82	
		1260	0.94	0.96	1.16	0.86	
Regression results by $h = \alpha N^{\beta}$							
α			0.2424	0.2597	0.2798	0.1379	
β			0.1928	0.1870	0.2024	0.2614	
r^2			0.9959	0.9932	0.9952	0.9899	
Test item		Unit	Analysis results				Specification
			1	2	3	4	
Wheel tracking evaluation	Dynamic stability	Passes/mm	15,750	14,400	12,600	15,750	KSF-2374: 2000
	Deformation rate	mm/minute	0.0027	0.0020	0.0033	0.0027	

Table 3.
Wheel tracking results for 19 mm modified mix specimens.

Marshall specimens were immersed in a temperature-controlled water bath at 60°C for 30 minutes, and then Marshall stability was tested for each specimen in the environmental chamber at 60°C.

A downward force was applied on each specimen at a speed of 50.8 mm/minute. The estimated Marshall stability (MS) and the corresponding Marshall breakdown energy were computed and are listed in **Table 4**. Likewise, the indirect tensile (IDT) test for each Marshall specimen at room temperature with a speed of 50.8 mm/minute was performed. The peak load and the IDT breakdown energy measured are also listed in **Table 4**.

3.3 Analysis of rut prediction by new mix design and Marshall stability

To validate the rut depth equations derived, Eq. (11) is adopted because it is relatively simple. Note that the rutting equation, Eq. (11), still contains the fatigue cracking variable W_{1d}, which belongs to the solid property. Traditionally, no rutting equation contains the solid fatigue cracking property (W_{1d}) in its expression. The existence of the solid property is a natural consequence because the asphalt mix after compaction, even at the rutting temperature, forms a solid structure, and breakdown appears with repeated loading [1]. Of course, the fatigue cracking has solid property, too, as shown in Eq. (6). The traditional empirical rutting equation often described by $h = \alpha N^{\beta}$ (where h is the rut depth, N is the number of wheel passes, and α and β are physical parameters) does not possess the solid property, in its expression.

The rut depth at 1260 cycles in **Tables 2** and **3** (the small wheel tracking number after the post compaction) with respect to W_{1d} from **Table 4** are selected to regress Eq. (11). Additionally, a relatively small N (=1260 cycle) and nearly the same β make the term h_N/N^{β} become almost h_N because N^{β} is nearly constant.

The excellent curve-fitting result shown in **Figure 2** is regarded to be a great surprise under all assumptions made to come up with Eq. (11). **Figure 2** clearly validates the legitimacy of both rutting and fatigue cracking assumptions to formulate Eq. (11), although all hypotheses made in driving Eq. (6), Eq. (7), Eq. (8), Eq. (9), Eq. (10), and Eq. (11) need more extensive validation.

Even though only eight different asphalt mixes are chosen (four regular modified and four porous modified asphalt mix specimens), the accuracy shown in regression by Eq. (11) encourages the promising future of all performance equations derived. Even though more extensive validation may be needed in the future, the concept introduced seems to provide a reasonable scientific foundation.

The empirical Marshall stability (MS) data are correlated to the rut depth data at N = 1260 cycles to compare the prediction result of the scientifically derived Eq. (11).

The rut depth at N = 1260 cycles predicted by the MS for the four porous and the four regular modified asphalt mixes are demonstrated in **Figure 3**.

The figure shows poor rut prediction accuracy ($r^2 = 0.5048$) in the MS method compared to the scientifically derived rut equation, Eq. (11), ($r^2 = 0.9542$) as shown in **Figure 1**. This comparison clearly demonstrates a wise choice of performance equations derived for rutting and fatigue cracking. The empirical MS mix design method is still widely used in most of the European and Asian countries as a practical mix design method because no accurate design method is known so far and is relatively simple to use. Thus, despite considerable errors found in the empirical MS method, it is still around in the asphalt mix design area.

The peak IDT load is usually regarded as the stiffness of a given mix and is an alternative to the MS method. Its rut prediction is manifested in **Figure 4**, which shows a poor result ($r^2 = 0.6644$) like the MS method ($r^2 = 0.5048$) shown in **Figure 3**.

Kinds of mixes		Marshall stability (kgf)		Break-down energy in Marshall Stability (kgf-mm)	Maximum IDT load (kgf)		Break-down energy in IDT (kgf-mm)
		Measured	Average	Average	Measured	Average	Average
13 mm – surface course specimen in modified porous asphalt mixes	1	1166.0	1135	3076	1050.1	982	742.4
		1081.5			910.5		
		1157.5			986.0		
	2	1265.5	1161	1696	1189.0	1305	1093
		1155.0			1341.5		
		1063.0			1558.5		
	3	785.5	894	1991	980.5	973	1112
		906.5			1063.0		
		991.0			958.0		
	4	927.0	979	1903	991.0	930	839.1
		989.5			901.0		
		1021.0			876.5		
19 mm- surface course specimen in modified asphalt mixes	1	1626.5	1447	3774	1950.0	1761	1976
		1739.5			1593.0		
		1556.5			1956.5		
		1396.5			1739.5		
		1208.0			—		
		1156.5			—		
	2	1546.5	1657	3547	2151.0	2040	1786
		1835.5			1791.5		
		1845.0			2143.0		
		1577.5			2184.5		
		1573.0			—		
		1564.0			—		
	3	1890.5	1787	4042	2190.0	2068	1528
		1828.5			1960.0		
		1977.0			2090.0		
		1741.0			2152.5		
		1566.5			—		
		1718.0			—		
	4	1975.0	1702	7042	1841.0	1929	2174
		1777.0			1856.0		
		1355.0			2089.0		

Table 4.
Data Measured by MS and IDT.

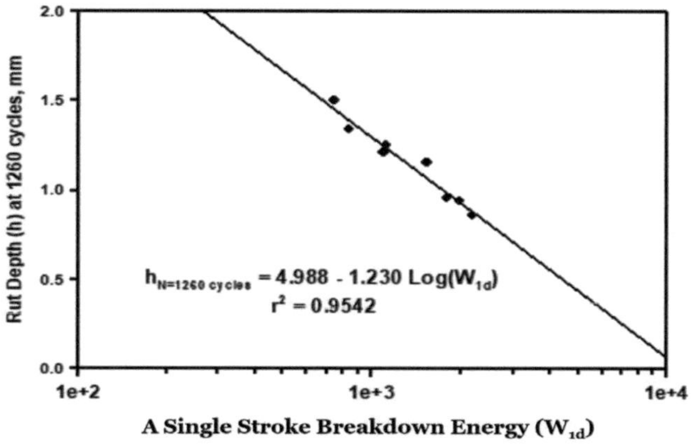

Figure 2.
Rut depth at N = 1260 cycles predicted by using Eq. (9).

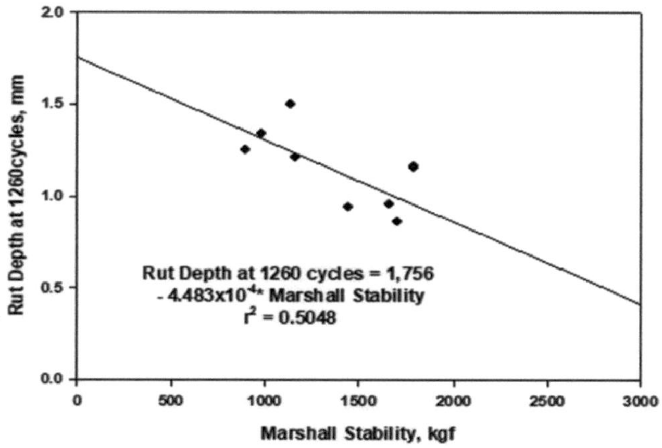

Figure 3.
Rut depth prediction at N = 1260 cycles by MS value.

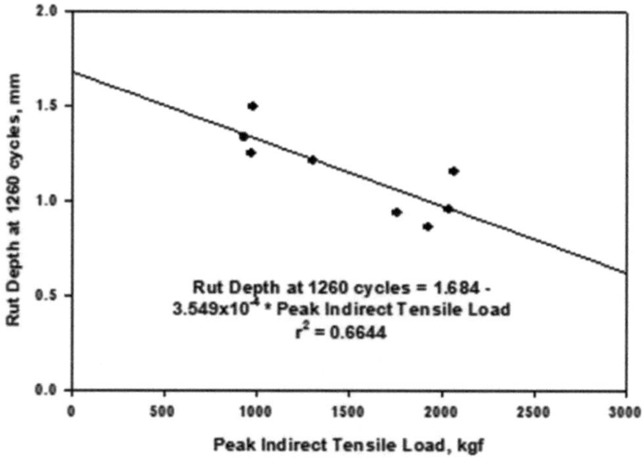

Figure 4.
Prediction of rut depth by the peak IDT load.

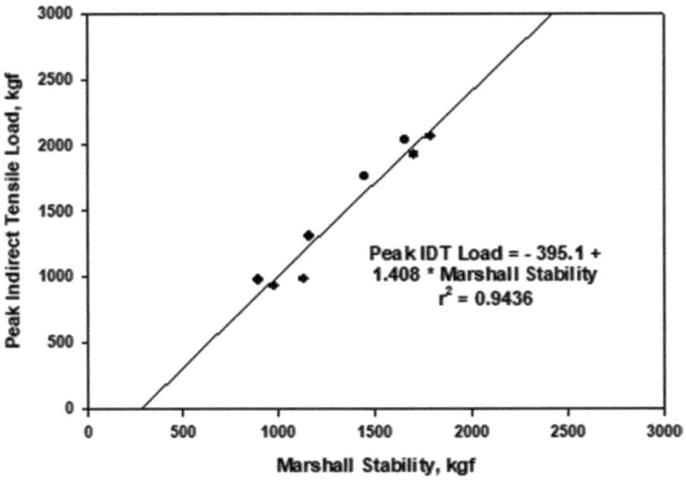

Figure 5.
Relationship between peak IDT load and MS.

Figure 5 exhibits below the relationship of the peak IDT at 25°C to the peak MS value at 60°C. The relatively good fitting (r^2 = 0.9436) between the two simply displays that the empirical MS value just represents the stiffness of the asphalt mixes such as the peak IDT representing the stiffness of the asphalt mixes.

4. Master performance equations for rigorous design equations

Eqs. (6) and (9) derived above represent asphalt mix performance equations (or the rigorous asphalt mix design equations), but they cannot take roll of master equations because they fail to provide constants in their equations. The reason is that constants P and R in Eq. (6) and constants E, F, C, D, j, and k in Eq. (9) are missing. With these constants specified, Eq. (6) for fatigue cracking and Eq. (9) for full rutting truly become the master performance (or the rigorous asphalt mix design) equations.

To estimate the constants (P and R) in the fatigue cracking equation (Eq. (6)), experiments should be performed to measure N_d (number of load repetitions reaching the fatigue cracking failure) and W_{1d} (single stroke breakdown energy) for a given asphalt mix. Obtaining a N_d requires the fatigue cracking experiment to perform repetitive loadings at a fixed loading weight and a speed for a given mix specimen to be broken down, and W_{1d} should be measured for the same mix specimen. Through many experiments for diverse asphalt mix specimens, many pairs of N_d and W_{1d} are available that can be regressed by using Eq. (6) to get P and R. By inserting these values into Eq. (6), one yields the master equation of the fatigue cracking (Eq. (6)). Once the master equation is formulated, estimation of W_{1d} for a certain asphalt mix can provide its fatigue cracking life (N_d).

This procedure can be applied identically in making the master equation of the full rutting equation (Eq. (9)). Here, the rut depth (h_N) is measured at the wheel-pass number at a damage (an inflection, or a boundary) point for a given rut mix specimen by using one of the lab wheel tracking devices. Note that the same W_{1d} obtained in the fatigue cracking experiments can be used for the same asphalt mix. After so many pairs of h_N and W_{1d} are obtained for many corresponding asphalt mixes, Eq. (9) is used to regress h_N with respect to W_{1d} to obtain constant values of E, F, C, D, j, and k.

These constants are inserted into Eq. (9) to formulate the master full rutting equation. If one measures a W_{1d} for a certain asphalt mix and it is put into the master full rutting equation, the rut depth can be predicted for the given mix.

These master performance equations can become rigorous hot mix asphalt design (or asphalt mix performance) equations.

Further attention is required for the W_{1d}. Once a W_{1d} is measured for a given asphalt mix will predict both the full rut depth and the fatigue cracking life simultaneously by using both corresponding master equations.

However, for the estimation of N_d by the fatigue cracking experiments and h_N by the full rutting experiments to formulate the master performance equations of Eq. (6) and Eq. (9), so many experimental works are required for measurements such as much effort of specimen preparation, set up of testing procedure, and tremendous amount of time consumption. Thus, another simple hot mix asphalt (HMA) design theory instead of formulating master performance equations will be suggested based on the performance equations of asphalt mixes in the next section.

5. A novel HMA design theory for the asphalt pavement construction and the asphalt mix quality check

The rigorous HMA design equations (or performance equations), Eq. (6) and Eq. (9), developed in the previous section are depicted schematically in **Figure 6**. It is noted that those design equations are the only function of the single stroke damage energy (W_{1d}). Each W_{1d} in the figure has a pair of data; one is the full rut depth (h_N), and the other is the fatigue cracking life (N_d).

W_{1d} can predict the fatigue cracking life (N_d) by using Eq. (6), and then Eq. (6) yields the full rut depth (h_N) in Eq. (9). This relationship is depicted mathematically in Eq. (12) below.

$$N_d = F_1(W_{1d}); h_N = F_2(N_d) = F_2[F_1(W_{1d})] = F_2(W_{1d}). \tag{12}$$

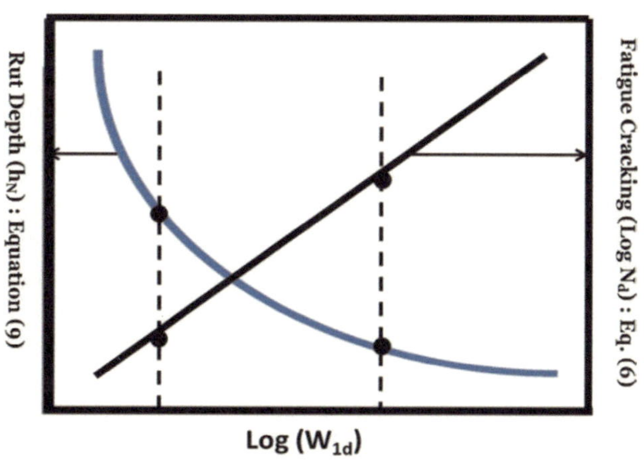

Figure 6.
Eqs. (6) and (9) are schematically drawn.

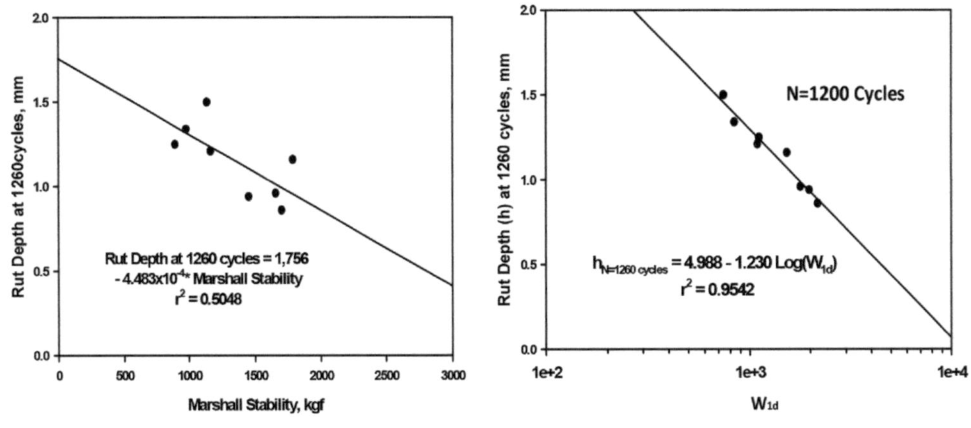

Figure 7.
Different accuracies are shown between MS and W_{1d}.

Figure 6 or Eq. (12) depicts that each single stroke damage energy (W_{1d}) determines both corresponding rutting and fatigue cracking value simultaneously. As indicated in **Figure 6**, the higher W_{1d} specifies the lower rut depth and the longer fatigue cracking life, and, reversely, the lower W_{1d} manifests the higher rut depth and the less fatigue life. That is, W_{1d} alone represents both performance values of rutting and fatigue cracking. Thus, the comparison of W_{1d} among different asphalt mixes is enough in finding different performance standings instead of evaluating rutting (Eq. (9)) and fatigue cracking (Eq. (6)) separatory. The direct use of W_{1d} in place of using Eqs. (6) and (9) reduces tremendous amount of experimental works in formulating master equations. This yields W_{1d} to be the new HMA mix design value. Thus, an innovative mix design method can be built upon W_{1d} measured for comparison among different asphalt mixes. In addition, W_{1d} can be used as the quality check of a new asphalt mix invented among others.

Measurement of W_{1d} for a given asphalt mix and then comparing it to other W_{1d} values to find its relative performance standing will give rise to the innovative, simple HMA mix design method. It is genuinely called the scientific, performance-based, HMA design method because W_{1d} itself represents the performance of rutting and fatigue cracking equations.

Figure 7 reiterates the different accuracies between the empirical Marshall Stability (MS) method and the single stroke damage energy (W_{1d}) method. In the asphalt mix design method for rutting as well as fatigue cracking performances, the W_{1d} method should be preferred over the MS method.

Since W_{1d} (or N_d) is a function of temperature, measured values will vary with temperature, especially if the material is changed. Asphalt mix materials (containing straight or modified asphalt binders) can undergo local crystallization or glass transition at low temperatures. Thus, the following temperatures should be considered for low-temperature measurement of W_{1d}. At these temperatures, an asphalt mix design criteria must be set differently.

1. 15°C (no phase change occurs from 10–70°C).

   Evaluation of rutting or fatigue cracking resistance simultaneously by measuring W_{1d} for all asphalt concrete mixes at 15°C.

2. −15°C (some molecules in asphalt binders undergo local crystallization).

   The W_{1d} of all asphalt concrete mixes should be measured to assess crack resistance.

3. −30°C (some molecules in the asphalt binder underwent a glass transition).

The W_{1d} of all asphalt concrete mixes should be measured to assess low-temperature crack resistance.

6. Conclusions

In the previous study, Huh [3], the full rutting performance equation (Eq. (1)) has been derived as a function of only a damage wheel pass number (N_d) that is hard to measure. It is explained that N_d in Eq. (1) is set to be identical to the fatigue cracking repetition number at failure (N_f) in the fatigue cracking study.

A fatigue cracking model in the literature is adopted and modified to provide a new variable easily measurable in the laboratory. It is known to be the single stroke damage energy (W_{1d}) that changes N_d (hard to measure) in the full rutting equation into W_{1d} (easily measurable).

Change of the variable constitutes two performance equations (the full rutting equation and the modified fatigue cracking equation). These two equations turned out to become rigorous mix design equations. To use them in practice, the master equations (specifying constants in two performance equations) must be formulated accordingly. The formulation requires tremendous amount of experimental works.

It is noted that two master performance equations display as a function of a single variable (W_{1d}) that controls both performance equations simultaneously by one single value. Thus, instead of using both master performance equations as rigorous mix design equations, direct comparison of the single variable (W_{1d}) can do the same role of using two equations.

Thus, measuring W_{1d} and comparing it to all others will constitute an innovative, simple, accurate, revolutionary, performance-based asphalt mix design method dealing with both regular and recycled asphalt mixes. This is the extraordinary discovery in the hot mix asphalt design method proposed for the first time.

Note that W_{1d} itself for a particular mix represents mix compositions, binder types, aggregate gradations, and environmental factors influencing the performance of mixes. Accurate mix design can save a tremendous amount of construction budgets due to avoiding early pavement damage or undesirable mix design criteria. Depending on the mix design value, a particular asphalt pavement life could be either 6 or 20 years. W_{1d} determines the life expectancy. That is why the proper mix design method is important.

Current mix design methods, Marshall Stability, Gyratory Compactor, Volumetric Design, Rice method, Balanced Mix Design, Vheem's Design, Semi-Circular Bending Test (SCB), and Disc-shaped Compact Tension test (DCT) are available in the industry. All these are either empirical, contain considerable errors, or fail to relate their design values to asphalt pavement performances. The scientifically proposed asphalt mix design method proposed in this study could be the best alternative to solve problems involved in all design methods mentioned.

Development of a new asphalt mix design method like measurement of one-stroke damage energy (W_{1d}) for all kinds of asphalt mixes, choosing a measuring equipment (a single equipment for all performance temperatures), specimen preparations, procedures of measurement, analysis of measured data, and overall setup of mix design criteria require further research support for actual implementation in the industry.

Author details

Jung Huh
GreenInnoTechUSA, Salt Lake, Utah, USA

*Address all correspondence to: jdhuh4712@gmail.com

References

[1] Gillespie T, Settineri WJ. The effect of capillary liquid on the force of adhesion between spherical solid particles. Journal of Colloid and Interface Science. 1967; **24**:199-202

[2] Mehrotra VP, Sastry VS. Pendular bond strength between unequal-sized spherical particles. Powder Technology. 1980;**25**:203-214

[3] Huh JD. Development of a novel scientific hot mix asphalt design method based on asphalt performance equations. Journal of Construction and Building Materials. 2023;**2023**

[4] Finn F et al. The use of distress prediction subsystems for the design of pavement structures. In: Proceedings, 4th International Conference on the Structural Design of Asphalt Pavements. Ann Arbor, MI: University of Michigan; 1977

[5] Dijk WV. Practical fatigue characterization of bituminous mixes. Journal of Association of Asphalt Paving Technologists. 1975;**44**:38-74

[6] Dijk WV, Visser W. The energy approach to fatigue for pavement design. Association of Asphalt Paving Technologists. 1977;**46**:1-40

[7] Abhirup BR-C, Mofreh S, Miguel M-G. Proceedings of the 9th International Conference on Maintenance and Rehabilitation of Pavements-Mairepav 9. Springer Nature; 2020. pp. 687-698

Chapter 5

Self-Healing Asphalt Using Steel Fibers: Induction Heating

Muhammad Aniq Gul and Muhammad Mudassir Ahmad Alwi

Abstract

Asphalt pavements are essential to our transportation infrastructure, but they have a significant drawback: they require frequent maintenance. However, innovation is revolutionizing the field of materials engineering and could change the way we think about road construction and maintenance. Incorporating self-healing technology into the road design phase holds the promise of extending the durability of roads, eliminating the need for frequent maintenance, leading to significant time and cost savings. Past research has shown that asphalt pavement can restore its stiffness and strength during periods of rest due to the closure of micro-cracks that naturally occur under the stress of traffic loads. Recently, interest in utilizing induction heating to improve the self-healing characteristics of asphalt pavements is revived. Integrating self-healing technology into the road design process holds the promise of revolutionizing road construction and maintenance procedures. Self-healing asphalt technology harnesses asphalt's natural ability to repair itself. By delaying the need for traditional replacement of heavily damaged roads, this technology reduces the production of asphalt mixtures and also cuts down on energy usage during replacement reducing the environmental impact of waste disposal. This chapter provides and insight to the three most promising options and outlines the future development in self-healing asphalt technology.

Keywords: asphalt pavements, self-healing, sustainable materials, smart pavement technologies, materials engineering, induction

1. Introduction

Highways play an integral role in our daily lives, serving as essential conduits for commuting to work, accessing services, engaging in social and recreational activities, and various other pursuits. The global landscape features an extensive network of public roads, covering thousands of kilometers and incorporating both concrete and asphalt pavements. The utilization of pavements in transportation infrastructure presents a distinctive challenge for achieving sustainability due to the extensive reliance on non-renewable materials during construction and maintenance, resulting in significant environmental impacts.

With the growing awareness of environmental sustainability in transportation infrastructure, both public and private contractors are increasingly acknowledging the importance of integrating sustainable products, processes, and technologies.

 IntechOpen

This recognition is evident across various stages, including the design, construction, operation, and maintenance of highways and airports, with a specific emphasis on pavements. While sustainability considerations have been addressed in the past, they were often indirect or informal. However, there is currently a notable increase in efforts to quantitatively assess the sustainability impacts of pavements and systematically incorporate them into decision-making processes [1].

Traditionally, economic factors dominated decision-making in pavement-related considerations. However, in recent years, environmental and social components have gained increased significance. The first step in setting standards and assessing progress is to measure sustainability. It is essential to identify where environmental impacts occur throughout the life cycle of pavements and evaluate how different sustainability initiatives can reduce these impacts. This understanding helps in making informed decisions regarding pavement-related matters. The pavement life cycle, illustrated in **Figure 1**, provides a helpful framework for depicting the various stages a pavement undergoes, starting from the initial design development and extending to the conclusion of its functional life.

The incorporation of sustainable materials in pavement construction significantly influences various stages, including design, building, and maintenance/preservation, leading to the extension of a specific asphalt pavement's lifespan. Numerous sustainable techniques have emerged, involving enhanced or innovative designs and the utilization of recycled materials and industry byproducts. For instance, the development of long-lasting pavements aims to enhance sustainability by promoting extended service lives, reduced maintenance needs, fewer repairs, and minimized traffic disturbances.

The process of constructing, repairing, and maintaining roadway pavements involves substantial amounts of construction materials, which undergo extraction, processing, transportation, manufacturing, and placement. Creating an enhanced pavement characterized by durability, cost-effectiveness, eco-efficiency, and superior performance is pivotal for sustainable infrastructure. Various strategies in pavement construction focus on minimizing energy consumption and integrating recycled materials and industry byproducts to strengthen sustainability.

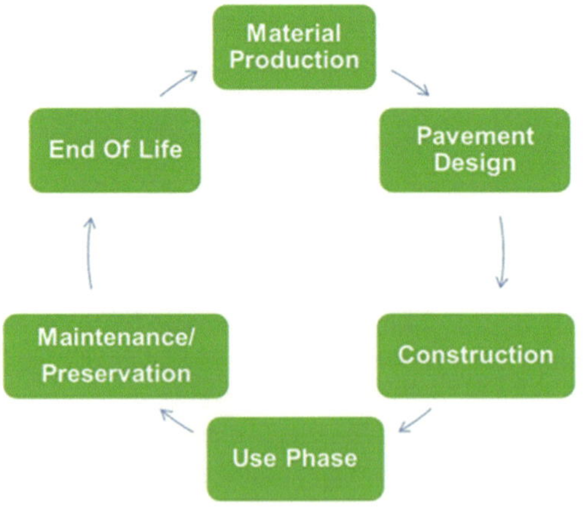

Figure 1.
Pavement life-cycle stages [2].

Figure 2.
Example of pavement damage, fatigue cracking.

The Federal Highway Administration (FHWA) has shifted its emphasis toward preserving and addressing the deterioration of the nation's highways. Unlike a predominant focus on rehabilitation, preventive maintenance treatments aim to refresh surfaces and address functional signs of pavement deterioration, such as minor cracking, friction issues, or asphalt pavement oxidation, without necessarily tackling structural problems. Preventive maintenance plays a crucial role in preventing minor deterioration, delaying pavement failures, and reducing the necessity for corrective maintenance or rehabilitation, ultimately extending the pavement's lifespan. When choosing pavement repair solutions, it is imperative to consider the economic and environmental impacts of various pavement maintenance and preservation operations. Numerous studies have been conducted to assess the cost-effectiveness of pavement preservation through life cycle cost analysis (LCCA) (**Figure 2**) [3].

2. Background of smart asphalt pavement mixes

Asphalt pavement plays a crucial role in withstanding heavy traffic and diverse weather conditions, ensuring safe and comfortable driving conditions at a reasonable cost and with minimal environmental impact [4, 5]. Recently, there has been a growing emphasis on introducing new capabilities, particularly within the realm of civil engineering, specifically Transportation Engineering [1, 6, 7]. It is noteworthy that the asphalt binder commonly used in constructing road pavements is a byproduct of petroleum crude oil distillation. Considering the extensive coverage of road pavements supporting traffic and their proximity to populated areas, highways and parking lots often make up over 35% of metropolitan areas. Recognizing these factors, scientists in Transportation Engineering have been enhancing asphalt mixtures to equip them with additional capabilities, yielding various social, environmental, and financial benefits [8, 9]. This functionalization process has given rise to smart materials, which are defined as intelligent substances capable of responding to stimuli like temperature and stress.

Introducing new additives into asphalt mixtures presents an opportunity to enhance road safety, particularly in regions where winter conditions increase the risk of traffic accidents due to reduced friction between tires and pavement [10–12]. Traditional methods to address this challenge involve the use of deicing chemicals [13] and the integration of conductive elements into asphalt mixtures [12, 14]. Deicing chemicals operate through a chemical process to melt ice and snow [15, 16], while conductive elements rely on microwave or induction heating units for the same purpose [17, 18].

An alternative approach involves the use of superhydrophobic asphalt mixtures [19, 20]. These mixtures resist water droplets by coating the asphalt surface with nano/micro particles, reducing the formation of ice and snow [20–22]. This method contributes to safer road conditions during rainy and cold weather [20, 23]. Moreover, the surface coating facilitates self-cleaning, promoting the removal of dirt particles from the road surface. This self-cleaning capability is effective on three different types of surfaces:

- *Superhydrophobic:* Water droplets rolling on surfaces and carrying previously deposited dirt is a phenomenon recognized as the Lotus Flower effect. This effect describes the ability of surfaces to repel water and facilitate the self-cleaning process, like how water droplets roll off lotus leaves, carrying away dirt and contaminants with them [24].

- *Super hydrophilic*: During rainy periods, water droplets spread across the surface, facilitating the removal of adsorbed dirt and making it easier to clean.

- *Photocatalytic effect*: Photocatalytic materials possess the capability to break down organic pollutants, such as oils and greases, through a degradation process.

Consequently, the incorporation of a self-cleaning feature in asphalt mixtures represents a promising avenue for reducing accidents in areas affected by oil spills and dust [4]. Beyond its potential impact on road safety, intelligent asphalt blends present an opportunity to address broader environmental challenges, particularly in mitigating air pollution and alleviating the Urban Heat Island (UHI) effect.

Air pollution has emerged as a significant environmental concern in major cities [25–27], drawing global attention through climate change conferences and extensive media coverage that underscores its adverse effects on human health. Recognizing the imperative for eco-friendly solutions, there is a growing awareness of the influential role that traffic on roadways, including automobiles [28, 29], and the road surfaces themselves play in reducing pollution.

Road surfaces commonly constructed with asphalt mixtures can undergo improvement by introducing photo-catalysts to aid in air purification through pollutant photo-oxidation [4, 30–32]. When exposed to ultraviolet (UV) light and humidity, semiconductor oxides like TiO_2 and ZnO engage in a process that oxidizes pollutants such as NOx and SO_2 while simultaneously breaking down organic molecules [4, 22, 31]. Researchers are actively exploring the incorporation of semiconductor nano/microparticles in a coating on asphalt mixes to harness their photocatalytic capabilities, effectively reducing air pollution levels [8, 33]. This innovative approach shows promise in advancing environmental sustainability within civil engineering practices. Urban heat islands (UHI) are characterized by significantly higher temperatures in urban areas compared to their rural surroundings due to human activities.

Thermochromic asphalt mixtures offer a solution by incorporating materials with higher mechanical strength and resilience to aging through color alteration [34, 35]. Addressing the high surface temperature caused by the traditional black color of asphalt roads, which absorb a significant proportion of sunlight, can enhance material performance [36, 37].

Another approach to mitigate urban heat island (UHI) effects involves utilizing phase change materials (PCM) to incorporate latent heat thermal energy storage (LHTS) capabilities [38–41]. This utilization of PCM leads to reduced temperature fluctuations, enhancing the mechanical properties of asphalt mixtures, which, in turn, minimizes rutting and prevents thermal cracks [4, 41]. To address the common pavement issue of cracking, particularly during the microcracking phase, researchers are exploring methods to improve self-healing capabilities. While asphalt materials naturally mend during rest periods, ongoing research focuses on innovative approaches such as incorporating conductive materials, nanoparticles, ionomers, and maltenes-rich microcapsules to maximize this self-healing capability [4, 42].

In the past decade, there has been a noticeable increase in research efforts dedicated to advancing the functionalities of asphalt mixtures, often referred to as smart asphalt. This category encompasses a range of features, including but not limited to photocatalytic, self-cleaning, self-healing, superhydrophobic, thermochromic, deicing/anti-icing, and latent heat thermal energy storage characteristics [4, 8, 12, 42, 43]. It is crucial to highlight emerging trends in materials, technologies, and processes that require thorough examination. These trends include the comparative analysis between asphalt pavement and asphalt bitumen materials [44], the nuances of pavement management [45], the intricate systems of permeable pavement [46], and a comprehensive exploration of pavement distress [47, 48].

Within this landscape, a subset of researchers has directed their focus specifically toward the intricate realm of smart asphalt mixtures, delving into the nuanced investigation of their self-healing properties [49]. This focal point indicates a burgeoning interest in the exploration and comprehension of the diverse capabilities that can be imbued within asphalt mixtures to augment their overall performance and sustainability. This trajectory of research reflects a collective endeavor within the civil engineering domain to unlock the full potential of smart asphalt and propel advancements in infrastructure materials and technologies.

A diverse array of materials and their corresponding healing mechanisms, along with references and publication years, is summarized in **Table 1**. In 2012, Jin et al. [58] explored polymer healing through the encapsulation of healing agents. For concrete, Jonkers and Schlangen [53] investigated a bacteria-based healing mechanism in 2009. Additionally, Dry [54] introduced a Hollow Fiber healing approach in 1996, Boh and Sumiga [56] explored Microencapsulation in 2008, and Kishi et al. [57] delved into the use of expansive agents and mineral admixtures in 2007. In the realm of asphalt, Tabatabee and Shafiee [51] researched nanoparticles as a healing mechanism in 2012. Garcia et al. [52] employed steel fiber—induction heating for asphalt healing in 2011, while Su et al. [50] investigated rejuvenator encapsulation in 2013. Coatings, as explored by Wilson and Magnus [59] in 2011, involved healing agent (resin) encapsulation. In the realm of composites, Kirkby et al. [60] focused on Memory Alloys in 2008. Lastly, Lumley [61] investigated Press and Sinter Powder Metallurgy as a healing mechanism for metals and alloys in 2007. This summary provides an overview of the materials, their healing mechanisms, references, and respective publication years, offering a comprehensive understanding of the research landscape.

Material	Healing mechanism	Reference	Year
Asphalt	Rejuvenator encapsulation	Su et al. [50]	2013
	Nanoparticles	Tabatabee and Shafiee [51]	2012
	Steel fibers—induction heating	Garcia et al. [52]	2011
Concrete	Bacteria	Jonkers and Schlangen [53]	2009
	Hollow fiber	Dry [54]; cited in [55]	1996
	Microencapsulation	Boh and Sumiga [56]; cited in [55]	2008
	Expansive agents and mineral admixtures	Kishi et al. [57]; cited in [55]	2007
Polymer	Healing agent encapsulation	Jin et al. [58]	2012
Coatings	Healing agent (resin) encapsulation	Wilson and Magnus [59]	2011
Composites	Memory alloys	Kirkby et al. [60]	2008
Metals and alloys	Press and sinter powder metallurgy	Lumley [61]	2007

Table 1.
Materials that have effectively incorporated self-healing technology.

3. Self-healing technologies for asphalt pavements

Over the past decade, remarkable progress in self-healing technology has significantly influenced the design of asphalt pavement [62–70]. This innovative technology introduces an alternative strategy for road maintenance by incorporating an internal healing system to tackle damage. The main aim of self-healing technology is to enhance the recovery process of material systems following damage, be it on a local or global scale. Its objective is to mitigate local or global damage while concurrently prolonging or restoring the functionality and lifespan of the affected part, system, or device. Asphalt pavement is a self-healing material. When subjected to rest periods, asphalt pavement has the potential to restore its stiffness and strength by closing the microcracks that occur when the pavement is subjected to traffic loads.

The purpose of self-healing technology is to facilitate the recovery of material systems after experiencing damage. Its goal is to minimize the extent of damage and prolong or rejuvenate the functionality and lifespan of the affected part [71]. Fisher provides a definition of self-healing and self-repair for a material or system, stating that it involves the capacity to largely revert to an initial, proper operating state or condition after exposure to a dynamic environment. This is achieved by making necessary adjustments to restore normality and/or resisting the formation of irregularities and/or defects [71].

In an asphalt pavement system, the process of repairing cracks takes place through the wetting and inter-diffusion of materials across the two faces of a microcrack. This allows the pavement to regain the properties of its original material [71, 72]. The autonomous asphalt self-healing process comprises three key steps. Firstly, it involves the wetting of the two faces of a microcrack. Following this, there is the diffusion of molecules from one face to the other. Finally, the diffused molecules undergo randomization to achieve a strength level equivalent to that of the original material. These sequential stages demonstrate the systematic nature of the self-healing mechanism within asphalt, contributing to its enhanced durability and structural integrity (**Table 2**).

Factors influencing self-healing		
Binder properties	**Asphalt mix properties**	**Environmental conditions**
Binder type	Binder content	Temperature
Chemical composition		
Viscoelastic properties	Aggregate structure and gradation	Loading history
Surface free energy		
Aging	Thickness	Rest period
Diffusion		

Table 2.
Factors that influence the healing of asphalt concrete [73].

Until now, researchers have examined two distinct self-healing technologies for asphalt pavements: (a) rejuvenator encapsulation and (b) induction heating [62–70]. These technologies offer innovative approaches to address damage in asphalt pavements. Rejuvenator encapsulation involves encapsulating a rejuvenating agent within the pavement material, which is released when cracks occur, aiding in the healing process. On the other hand, induction heating utilizes electromagnetic energy to heat the asphalt, allowing it to flow into and fill cracks, restoring the pavement's integrity. These technologies show promising potential to enhance the durability and longevity of asphalt pavements, reducing the need for frequent repairs and maintenance. The exploration of these self-healing technologies reflects a commitment within the engineering community to continually enhance the durability and performance of asphalt pavements.

3.1 Rejuvenator encapsulation

The encapsulation of a rejuvenator has emerged as a preferred method for self-healing, offering a way to restore an aged binder to its original chemical, physical, and mechanical properties. Research suggests the development of various capsule types containing rejuvenators that demonstrate both thermal and mechanical resilience, ensuring they can withstand the asphalt production process [50, 52, 63, 67–69, 74, 75]. This approach shows promise in enhancing the effectiveness of asphalt pavements, providing a sustainable solution to rejuvenate aged binders and improve overall pavement performance. This represents a promising direction for integrating rejuvenator encapsulation into asphalt technologies, showcasing its potential to improve the overall performance and lifespan of asphalt pavements. However, one drawback of this approach is the significant quantity of microcapsules needed to make the process effective. Despite facing challenges, ongoing research endeavors aim to optimize the use of microcapsules to maximize the benefits of rejuvenator encapsulation in asphalt pavements, addressing the need for more sustainable and durable road surfaces.

However, incorporating an excessive quantity of microcapsules into the asphalt mix may compromise pavement quality, potentially leading to premature pavement deterioration. Studies conducted by Garcia et al. [52] and Sun et al. [75] observed a reduction in stiffness when microcapsules were applied to asphalt. This highlights the importance of carefully optimizing the dosage of microcapsules to maintain the desired pavement characteristics and structural integrity, ensuring the longevity and performance of the road surface. They explained that the release of the rejuvenator caused the weakening of the asphalt binder, leading to a reduction in viscosity. It underscores the significance of a balanced approach in implementing rejuvenator encapsulation for effective pavement enhancement (**Figure 3**).

The encapsulation process for rejuvenators at various stages is depicted in **Figure 4**.

a. Cracks develop in the pavement due to various factors such as heavy traffic and environmental conditions.

b. Multi-cavity capsules filled with rejuvenators are inserted into the cracked mixture, causing a partial release of the rejuvenator within the crack zone.

c. The rejuvenator is gradually released from the capsules, permeating into the cracks and rejuvenating the damaged bitumen properties.

d. As a result, microcracks heal, and the bitumen is revitalized, restoring the pavement's integrity.

This visualization highlights the sequence of events involved in utilizing encapsulated rejuvenators to address pavement cracks, showcasing the potential for enhancing asphalt pavement resilience and durability.

Nonetheless, research findings [77, 78] highlight that the inclusion of sand granulates can lead to deformation in asphalt mixes. It is likely that the incorporation of microcapsules, resembling sand particles, has also played a role in the increased deformation of asphalt mixes, particularly contributing to rutting. Moreover, the chemical compounds used in the production of microcapsules, such as melamine-formaldehyde [79], might pose an environmental risk due to the potential for leaching. These aspects underscore the importance of carefully considering the impact of various additives on asphalt performance and the environment.

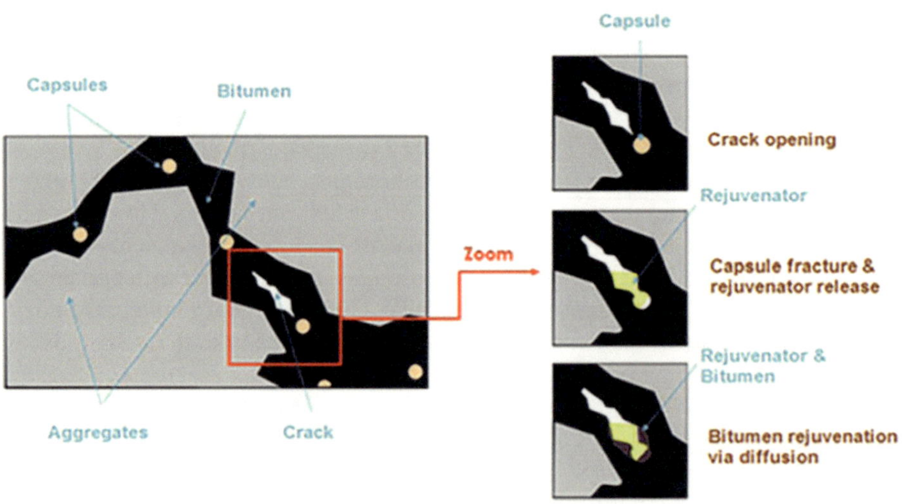

Figure 3.
Encapsulation of rejuvenator: a self-healing mechanism in asphalt mixtures [76].

DOI: http://dx.doi.org/10.5772/intechopen.1005427

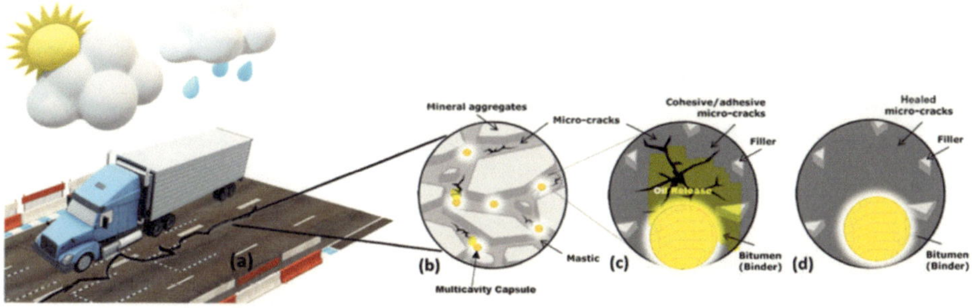

Figure 4.
Self-healing of asphalt with encapsulated rejuvenator at different levels [76].

Integrating microcapsules containing asphalt-recycling agents into asphalt proves to be a valuable method for enhancing its self-healing capabilities and mitigating the occurrence of brittle pavement cracking caused by asphalt aging. In practical engineering applications, it is essential for microcapsules to possess favorable thermal stability, mechanical properties, and consistent dispersion within asphalt or asphalt concrete. This is crucial for effectively addressing the self-healing of microcracks that develop as pavements undergo aging processes.

Several factors influence the self-healing performance of asphalt with embedded microcapsules. An examination of the thermal stability, mechanical properties, and dispersibility of microcapsules in asphalt was conducted through thermogravimetric analysis, nanoindentation tests, and fluorescence microscopy. Subsequently, a two-stage fatigue loading test was employed to investigate the impact of microcapsule content, temperature, time, degree of damage, and self-healing duration on the self-healing performance of asphalt-containing microcapsules. Various research studies revealed that the microcapsules exhibit satisfactory thermal stability, mechanical properties, and dispersibility. These resist thermal decomposition during asphalt concrete mixing, remain intact during the early stages of pavement life, and avoid agglomeration in the asphalt. The incorporation of microcapsules significantly enhances the self-healing performance of asphalt. The self-healing index of microcapsule-infused asphalt initially rises and then declines with increasing microcapsule content. Additionally, it increases with higher temperatures and longer durations but decreases with worsening damage, leading to an extended self-healing time (**Figure 5**) [81].

3.2 Electric arc furnace slag (EAFS)

Electric arc furnace slag (EAFS) exhibits favorable properties suitable for integration into construction materials, if the CaO and MgO content is carefully controlled to prevent potential material expansion and disintegration [80]. The effective combination of EAFS with bituminous compounds has been demonstrated, with specific instances of it carrying a CE marking, indicating compliance with EN 13043/AC as an aggregate for bituminous mixtures [82]. This certification assures that aggregates used in bituminous mixtures meet the required quality standards. Moreover, EAFS can be obtained at a comparable cost to other aggregates, ensuring that its extensive use does not compromise the economic feasibility of the mixture. This highlights the potential of EAFS as a valuable and cost-effective component in bituminous mixtures for construction applications.

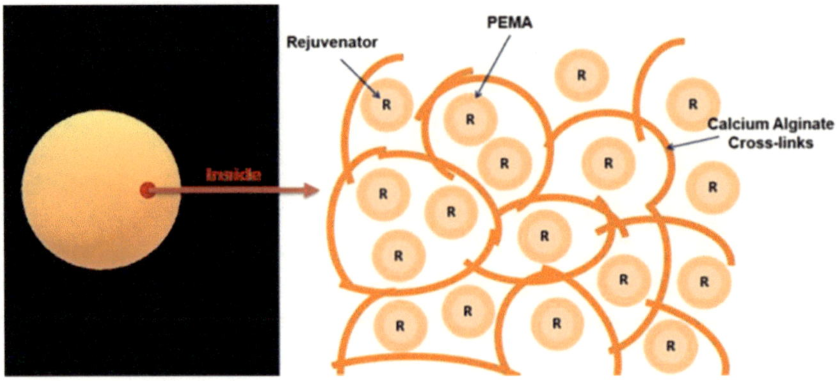

Figure 5.
Schemes of rejuvenator encapsulated by the calcium alginate crosslinks inside of a capsule [80].

Numerous benefits have been documented, with reports indicating that electric arc furnace slag (EAFS) surpasses quartzite gravel and diorite in terms of polished stone value (PSV) [83]. Studies by Asi [84] have further demonstrated a substantial improvement in skid resistance when 30% of limestone is replaced with EAFS in a bituminous mixture, as evaluated using a British Pendulum Tester. Both findings consistently recommend the use of EAFS as coarse aggregate for road surfaces to enhance skid resistance. Importantly, the introduction of EAFS minimally impacts the overall properties of bituminous mixes, including stability, water sensitivity, resilience modulus, and fatigue life [85].

It is noteworthy that various researchers have explored the application of microwave energy for heating bituminous mixes, presenting intriguing potential applications such as in-place surface mending, removal of damaged mixtures during patching, deicing frozen roadways, and substituting traditional energy sources like gasoline or gas in asphalt plants. Studies on in-place applications revealed that microwave penetration depth in a bituminous road surface can extend up to 15 cm at 4.5 GHz frequencies [86]. The research demonstrated that microwave energy, heating the mixture from within, does not burn the road surface. Researchers concurred and affirmed that bituminous mixes treated with microwave energy exhibited improved water sensitivity, making them more resistant to moisture damage [87]. These findings suggest the viability of using microwave heating for in-place fracture repair or in locations requiring layer removal due to ruts or potholes.

Similar conclusions were drawn regarding the production of bituminous mixtures, with microwave-heating technology showing the potential to reduce energy consumption and delay bitumen aging compared to conventional heating methods. Intriguing investigations have been conducted to examine the feasibility of incorporating microwave-heating technology in the manufacturing and recycling of bituminous mixtures [88]. However, researchers collectively acknowledge that microwaves have a limited capacity for heating bituminous mixes, influenced by the mineralogical features of the aggregates. For example, basalt or limestone tends to absorb more energy when exposed to microwave radiation than silica [87]. To enhance the susceptibility of bituminous mixes to microwave radiation, certain additives were employed. Some researchers used carbonyl iron powder [89, 90], while others supplemented the mixture with bituminous steel wool [91]. In both cases, there was a noticeable increase in susceptibility to microwave energy.

The studies incorporated various materials such as steel wool, scrap tire wire, silicon carbide, iron filings, and electric arc furnace slag (EAFS). Considering market costs and ease of industrial application, EAFS emerges as a promising alternative. Major European countries produce an estimated 8.5 million tons of EAFS annually [92]. Utilizing EAFS in bituminous mixtures for pavements and road surfaces not only addresses environmental concerns but has also led some waste-reduction companies to obtain a CE marking for steel slag aggregates, providing standardized approval for the use of slag in bituminous mixtures, like natural aggregates.

3.3 Self-healing by external energy source

The capacity of an asphalt mixture to naturally seal its microcracks from vehicle traffic during a resting period and in the absence of external loads is known as self-healing [93]. This occurrence is possible because the asphalt fills with microcracks, becoming fluid at higher temperatures. Capillarity between the crack walls further aids in this process [74, 94]. However, there is also the potential for artificially promoting healing, as demonstrated by numerous researchers in their studies [93, 95–98]. This approach has been extensively investigated over the last decade [99] and has emerged as a viable option for the recovery of microcracked asphalt pavement.

One method to induce healing in asphalt mixtures involves using an external energy source, such as electromagnetic microwaves, and incorporating particles that make the mixture responsive to heating, such as steel fibers, graphite, and steel slag [96, 100, 101]. When exposed to microwaves, these additives within the asphalt mixture heat up, raising the temperature of the asphalt, reducing its viscosity, and facilitating the natural filling of microcracks [93]. In the quest for self-healing asphalt pavements, researchers have explored innovative methods to trigger the mending process within asphalt mixtures. One promising technique involves the application of external energy sources like electromagnetic microwaves. These microwaves are utilized to activate specific additives within the asphalt mixture, such as steel fibers, graphite, and steel slag. Once exposed to microwaves, these additives heat up, causing the temperature of the asphalt to rise. As a result, the viscosity of the asphalt decreases, enabling it to naturally flow and fill in microcracks within the pavement structure. This process mimics the natural healing mechanism of asphalt, promoting enhanced durability and longevity of the pavement (**Figure 6**).

One laboratory approach to evaluating the healing capacity of asphalt mixes involves subjecting a sample to successive cycles of fatigue damage and healing. These cycles encompass specific steps: inducing fatigue damage to the sample, applying microwave heating to the sample, allowing the sample to rest, and subjecting it to further fatigue damage [97]. The data obtained from these cycles are instrumental in calculating the self-healing index (HI) of the mixture. For instance, if a sample undergoes a fatigue damage and healing cycle, the HI is determined by the ratio of the fatigue resistance exhibited immediately after the last test to the fatigue resistance after the first test. Fatigue resistance, in this context, relates to the number of load applications needed to reduce the sample's initial resilient modulus to a specific level [97, 102]. The HI, being variable, depends on several factors influencing asphalt mixtures, and severe fatigue damage can limit the mixture's self-healing potential, as evidenced by studies indicating that reducing the robust modulus from 50% to 60% diminishes the self-healing index from 1.33 to 0.24 [97].

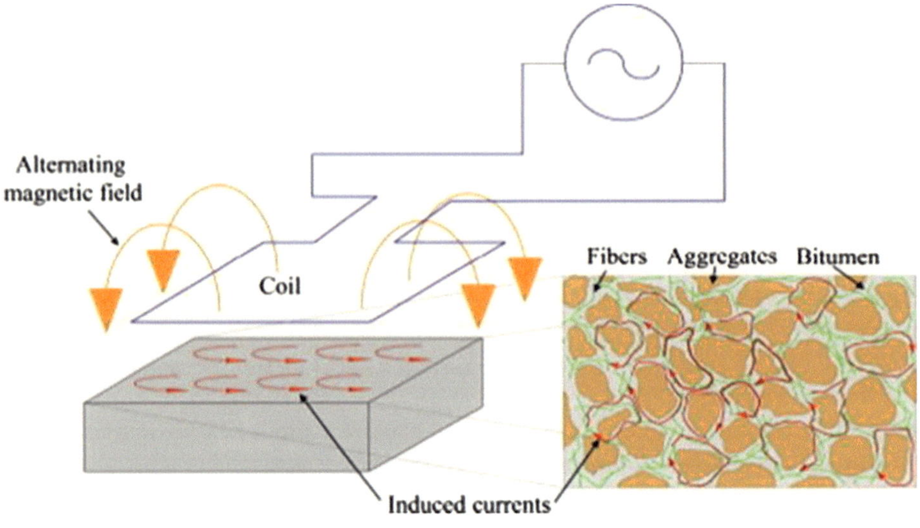

Figure 6.
A schematic diagram illustrating the fundamental principle of the induction heating method employed in self-healing asphalt [93].

Another crucial factor affecting the HI is the resting period, denoting the time between the first and second fatigue damage, where the specimen is free from load application. The optimal resting period for achieving a higher HI remains undetermined in the literature, with varying durations such as 2 hours [94], 3 hours [97, 103], 6 hours [104], and even periods exceeding 24 hours being explored [95, 102, 105, 106].

Additionally, the heating temperature significantly influences the self-healing capacity of asphalt mixes, and consensus on the ideal temperature is lacking. While there is a range where the HI increases proportionally with temperature, there exists a temperature limit beyond which the HI performance reverses [97]. Some studies propose that a heating temperature of 70–85°C enhances HI [107]. Notably, for stone mastic asphalt (SMA) mixtures, the literature suggests that the optimal temperature for a higher HI is 55°C [97]. Conversely, dense asphalt mixtures containing bitumen 70/100 may achieve a greater HI at a temperature of 55°C [102], according to some researchers (**Figure** 7).

Similarly, the self-healing capacity of asphalt mixtures is significantly influenced by the type and quantity of additives used. Noteworthy results were reported in a study where asphalt mixtures incorporated steel fibers with an average length of 10 mm, constituting 0.2% of the bulk of the asphalt mixture, resulting in outstanding self-healing index (HI) outcomes [108]. Another study, focusing on microwave-induced self-healing of asphalt mixtures [91], indicated that varying levels of steel fibers, specifically 0.2% and 0.4% per mass of the asphalt mixture, exhibited similar responsiveness to microwave radiation heating, characterized by equivalent heating rates. However, an observed concern during the addition of steel fibers to mixtures is the formation of clusters during mixing. To mitigate this issue, some researchers advocate for the use of 10 mm fibers [109], while others prefer shorter and thicker fibers [94]. The presence of clusters, aside from introducing heterogeneity to the asphalt mixture, can impact void volume and, consequently, the mechanical properties of the mixture. Addressing these concerns is vital for optimizing the self-healing potential of asphalt mixtures (**Figure** 8).

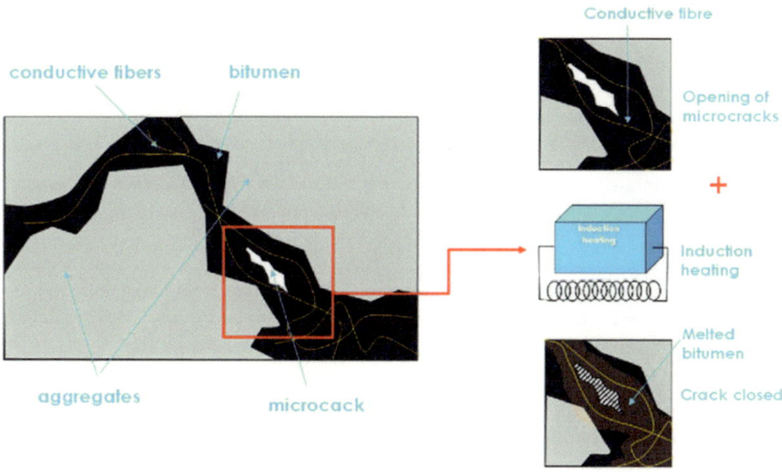

Figure 7.
The principle of induction heating [106].

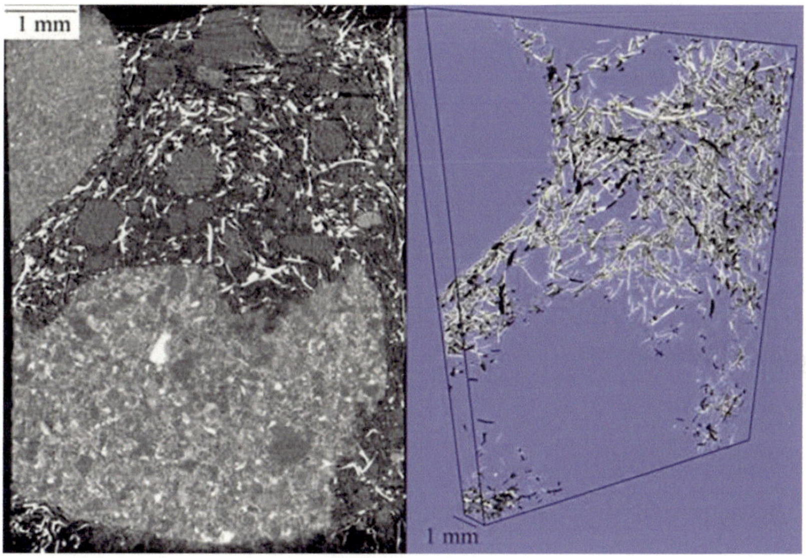

Figure 8.
Illustration of the steel fibers distribution in an asphalt concrete [110].

Reports indicate that the inclusion of steel slag in asphalt mixtures enhances their responsiveness to microwave heating, exhibiting more consistent temperature distribution than mixtures incorporating steel fibers [95]. Remarkably, this additive demonstrated outstanding mechanical performance in terms of mixture resistance, as evidenced by exceptional results in asphalt mixes [111]. Furthermore, studies have discovered that asphalt mixtures incorporating steel slag exhibit susceptibility to microwave heating, indicating its potential for effective self-healing properties [96]. In a separate study, excellent mechanical behavior results were achieved in mixtures containing 15% graphite powder, surpassing the performance of mixtures with 10% and 20% graphite powder content [73]. These findings highlight the versatility of

Rejuvenator encapsulation	
Advantage	The integration of rejuvenator encapsulation into asphalt technologies offers a promising avenue for enhancing the overall performance and durability of asphalt pavements. This innovative approach holds significant potential to bring about substantial improvements to the industry, ensuring that our roads and highways remain resilient and durable over time.
Disadvantage	To ensure effectiveness, a significant number of microcapsules is necessary, but this can negatively impact pavement quality and result in premature pavement deterioration.
Electric arc furnace slag	
Advantage	Rejuvenator encapsulation holds immense potential in revolutionizing asphalt technologies. By incorporating it, we can significantly enhance the performance and durability of asphalt pavements. It is a promising avenue that needs to be explored further to ensure the longevity of our roads.
Disadvantage	Maintaining appropriate levels of CaO and MgO is crucial to prevent any material growth or disintegration. By regulating these levels, we can ensure that the material remains stable and reliable, which ultimately leads to better performance and longevity.
Self-healing by external energy source	
Advantage	Over the last decade, the approach of using asphalt to recover microcracked pavement has been extensively researched and is now considered a practical option. This is because the asphalt could fill the tiny cracks and become more fluid at higher temperatures, enabling the repair process to take place.
Disadvantage	The healing index (HI) of asphalt mixtures cannot be assigned a fixed value due to the varying degree of healing and the impact of several factors. The self-healing potential of the mixture can be limited by severe damage caused by fatigue, for instance.

Table 3.
Comparison of the advantages and disadvantages of various processes for self-healing of asphalt.

different additives in influencing the self-healing characteristics of asphalt mixtures, showcasing the potential for optimizing performance based on specific material compositions (**Table 3**).

4. Use of self-healing technique

The introduction of self-healing technology marks a significant advancement. This innovation allows the road surface to undergo localized healing at critical points throughout its life cycle by incorporating fine steel fibers into the asphalt mix at these key locations. A mobile induction equipment, traversing the road surface, executes the healing process. A brief "heat-shot" is administered to the surface asphalt layer using steel fibers as a heat transfer medium. This action causes partial melting of the binder within the asphalt, effectively sealing any small fissures present in the asphalt mixture. Consequently, the asphalt mixture undergoes a "reset," initiating a new lifespan (**Figure 9**).

4.1 Inductive healing technique

The practice of heating an asphalt surface layer *in situ* has a long history, particularly with the widespread use of repave and remix techniques. In this method, infrared heaters were commonly employed, but the poor heat conductivity of asphalt, attributed to mineral aggregates, led to challenges. Excessive overheating

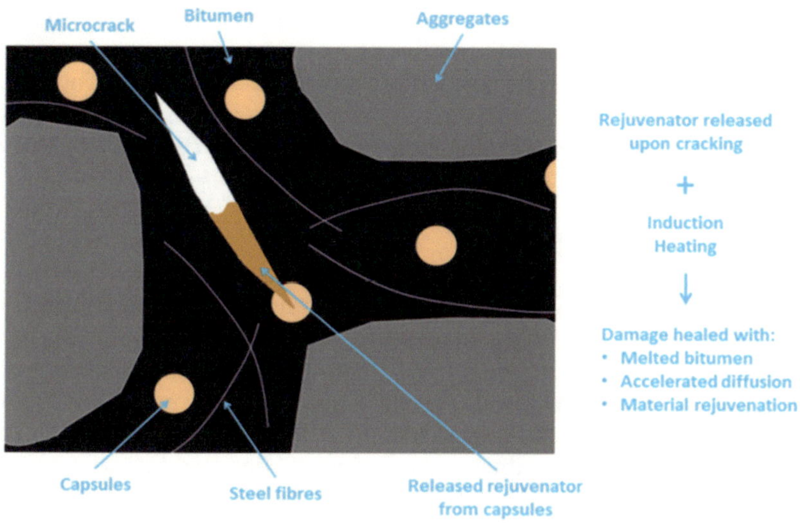

Figure 9.
Crack healing mechanism of the combined healing system [108].

was required to reach the desired temperature at the bottom of the asphalt layer, resulting in energy loss and frequent bitumen burning, evident in photos associated with blue smoke. This approach faced difficulties, especially in scenarios with moisture in porous asphalt layers, demanding strenuous efforts to achieve the necessary temperature.

In contrast, the newly developed induction technique offers a more direct solution. During asphalt mixture production, a small amount of highly refined steel filaments (or fibers) is incorporated. These steel fibers facilitate induction heating of the asphalt layer, responding to a rapidly changing magnetic field, akin to induction cooking plates in kitchens. This technique minimizes energy loss compared to gas burning.

The harmless and high-capacity induction technique, when applied to asphalt with steel fibers, results in their heating and transferring heat to the surrounding mastic. This enables intensive and uniform heating of the mastic across the entire thickness of the asphalt layer, causing momentary melting. As a result, fine cracks within the mastic close, resetting the asphalt to its original state. The advantage lies in the direct application of energy to the mastic, where fine hairline cracks form over time, ensuring minimal energy loss during the induction heat transfer. Only the effective part of the asphalt mixture requires heating.

4.2 Practical applications of self-healing asphalt pavements

Until this point, numerous field applications have already been implemented, and these can be categorized into various types of applications.

4.2.1 Single as well as double-layered porous asphalt

The self-healing technique proves versatile and applicable to both single-layered porous asphalt and double-layered porous asphalt, offering solutions to common challenges with the aim of prolonging pavement lifespan. There are ongoing efforts to apply self-healing technology to porous asphalt in scenarios where a shorter lifespan is expected, such as road bends with small radii or entry/exit ramps at intersections. Initial healing tests on a track have yielded promising results, showcasing the potential of the technology. While the surface can be heated as anticipated, quantifying the impact on service life remains a challenge, particularly since the road surface was only 3 years old during the assessment.

4.2.2 Silent joints (Brain Joint)

The term "Brain Joint" is derived from "Buried Reinforced Asphalt Interlayer Noiseless Joint," representing an innovative, noise-reducing joint suitable for specific applications. It serves as a silent alternative to traditional expansion joints, particularly ideal for locations like road bridges with asphalt surface layers, where eliminating pulse noise is essential. This innovative joint is particularly beneficial in areas near residential zones, where noise reduction is a priority. However, factors such as bridge span length exceeding specified limits may restrict the application of the Brain Joint in some cases. To overcome potential challenges, a specialized asphalt surface layer for the Brain Joint has been developed, featuring a self-healing design with incorporated steel fibers. This design enables the asphalt to heal any cracks caused by movements exceeding the bridge deck's design limits. The healing process is localized to the immediate vicinity of the bridge deck and can be performed more frequently, typically once every 2 years (**Figure 10**).

4.2.3 Implementing a two-layered porous asphalt solution for roads with lower traffic volume

Implementing a finely graded porous asphalt layer over a coarser porous asphalt layer yields sustained noise reduction benefits. In this design, rainwater efficiently drains through the thin top layer, then moves into the secondary porous layer, and eventually exits laterally through the coarser asphalt layer to the road's side. The pumping effect induced by traffic during rainfall enhances openness, ensuring the unclogged conditions required for effective noise reduction. This innovative approach suggests the potential for lifespans exceeding 10 years or more on highways (**Figure 11**).

Furthermore, the noise reduction capacity remains consistently effective over the lifespan, a notable difference from single-layered Dense Graded surfaces that quickly lose their noise reduction effectiveness due to congestion. The dense structure of a single-layered Dense Graded surface inhibits efficient sideways water drainage, leading to congestion and blockages. In the context of a two-layered porous asphalt, this issue appears to be alleviated, assuming there is effective water drainage toward the road's side. From the perspective of highway applications, ensuring unimpeded lateral water flow to the road's side proves to be a practical solution. Properly managing the unpaved road shoulder eliminates the need for drain gutters. However, careful consideration is necessary for situations where allowing free water flow to the side is impractical in the design. In cases of longer lengths, the use of drain gutters becomes necessary, albeit introducing challenges for maintenance (**Figure 12**).

The integration of self-healing technology emerges as a potential solution for road sections on the route that present challenges for the application of standard two-layered porous asphalt. This becomes especially pertinent in areas subjected to heavy

Figure 10.
Self-healing Brain Joint expansion joint [112].

Figure 11.
Example application of two-layer porous asphalt [112].

loads, such as side road connections, driveways, intersections, roundabouts, and similar locations. Consequently, deploying the self-healing technique has the potential to extend the use of two-layer porous asphalt to non-national (or lower-order) roads, contributing significantly to the reduction of environmentally unfriendly road traffic noise.

The practical application discussed above demonstrates the feasibility of utilizing self-healing asphalt, especially in high-traffic areas with a pronounced need for noise reduction. While initial experiments showcasing the healing action have been conducted, the economic viability of this technology is yet to be fully validated. The incorporation of costly steel fibers raises concerns about the economic justification of such applications. Recycling possibilities are crucial for future considerations, and given that the material remains thermoplastic, it can be recycled similarly to normal asphalt, ideally in a new self-healing pavement.

Figure 12.
Water draining ability of porous asphalt [113].

To optimize and validate this technique, an extensive investigation program is essential. This involves refining the conductive material added to the asphalt mix (currently steel fibers), determining the optimal quantity, and further developing the induction apparatus. Addressing these aspects will not only enhance the economic feasibility but also contribute to the broader implementation and success of self-healing technology in noise reduction for road surfaces.

5. Cost and environmental benefits of self-healing asphalt

The introduction of self-healing asphalt is expected to lead to longer intervals between road maintenance, consequently reducing traffic congestion and associated costs. To accurately assess the potential cost savings linked to the adoption of self-healing materials, it is recommended to compare the changes in material costs (increase) with the changes in maintenance costs (decrease). Additional expenses like operating costs, disposal fees, and environmental costs may be factored into the comprehensive cost-benefit analysis, depending on the specific application. For example, in the Netherlands, the combined annual savings related to significant repairs and costs associated with traffic congestion total approximately 82 million dollars for a 25% extension in the lifespan of asphalt and over 127 million dollars for a 50% extension across the entire porous asphalt pavement area [114]. Even if the price of self-healing asphalt was double that of standard bitumen, the Netherlands could save around 115 million dollars annually by investing in self-healing asphalt with a 50% extended lifespan, compared to traditional porous asphalt. Notably, the Netherlands, while relatively small in terms of European standards, contributes only 3% to the total European asphalt production, with just 1% used in surface layers [115]. Extrapolating the potential savings from the Netherlands to the entire European Union could result in a substantial sum, reaching up to 12 billion dollars [115].

Nevertheless, comprehending the complete advantages of self-healing technology in asphalt pavement design necessitates knowledge of the entire life cycle costs of asphalt pavements, encompassing financial, environmental, and societal aspects. Butt et al. [116] conducted a study examining the impact of self-healing on the lifespan, energy consumption, and environmental impact of asphalt pavements. Employing a life cycle analysis (LCA) framework in tandem with a numerical model simulating the self-healing capability of asphalt pavements [117], they determined that asphalt pavements with self-healing capacity increased the pavement's lifespan by 10% (from 20 to 22 years) compared to pavements lacking such capacity. This extension in lifespan corresponded to a 3% reduction in energy consumption and a 3% decrease in CO_2 emissions (1.5 tonnes). If the projected increase in the lifespan of an asphalt pavement reaches 100% (from 20 to 40 years, assuming self-healing technology can double the lifespan), the benefits in terms of cost reduction, reduced energy consumption, and diminished CO_2 emissions will scale proportionally [117].

In the past, there has been some research conducted to determine the carbon footprint and costs of both strategies of road maintenance over the years 1971–2015. For the several stages of construction and maintenance, we determined the specific capital and operational costs for the demanded commodities (see **Tables 4** and 5). Regarding the total costs, self-healing roads are 28% lower cost (capital expenditure from construction activities) and 36% lower cost (operational costs) compared to conventional roads. Together, the total costs of self-healing roads are almost 32% lower compared to the total cost of conventional roads. In the case of the traditional road, a decade following the initial construction of the infrastructure, a thin maintenance treatment known as a slurry seal is applied to the pavement surface, establishing a new wearing surface. Subsequently, milling and replacement of deteriorated asphalt become necessary due to the ongoing wear and tear of the pavement.

As evident from both case studies [118], the construction machinery sector stands out with the highest input costs, particularly during the construction year and recycling phases. Following is the road freight forwarding sector, which becomes more significant during road demolition, necessitating the removal and disposal of materials in a landfill. The bitumen industry, deriving from crude oil, a fossil fuel, also incurs substantial input costs. Aggregate input costs gain prominence solely during road construction, and specifically, only steel slag is used for self-healing roads. The peak input costs occur in the initial construction year for both cases. However, when comparing construction and maintenance costs, maintaining each type of road involves higher input costs, particularly for conventional roads compared to

Industry sector	Year 1	Year 10	Year 15	Year 23	Year 30	Total
	Construction	Slurry seal treatment	Milling and replacement	Recycling	Demolition and rebuilt	
Steel	$0	$0	$0	$0	$0	$0
Gravel	$49,096	$593	$7794	$13,181	$0	$70,664
Fuel oil	$23,418	$344	$9324	$18,499	$6406	$57,992
Bitumen	$75,336	$1882	$25,977	$44,283	$0	$147,477
Construction machinery	$65,184	$75	$38,427	$72,047	$22,617	$198,350
Electricity supply	$1	$0	$1	$1	$0	$3
Road freight forwarding	$27,539	$509	$17,195	$32,191	$70,560	$147,994
Total	$240,574	$3403	$98,718	$180,203	$99,583	$622,481

Table 4.
Overview of input costs of 1 km lane of conventional road [118].

Industry sector	Year 1	Year 8	Year 15	Year 21	Year 29	Year 36	Year 42	Total
	Construction	Microwave healing treatment		Milling and replacement	Microwave healing treatment		Demolition and rebuilt	
Steel	$574	$0	$0	$574	$0	$0	$0	$1149
Gravel	$47,947	$0	$0	$7794	$0	$0	$0	$55,741
Fuel oil	$23,418	$20,177	$20,177	$9324	$20,177	$20,177	$6406	$119,856
Bitumen	$75,336	$0	$0	$25,977	$0	$0	$0	$101,313
Construction machinery	$65,184	$10,500	$10,500	$38,427	$10,500	$10,500	$22,617	$168,228
Electricity supply	$1	$0	$0	$1	$0	$0	$0	$2
Road freight forwarding	$27,539	$0	$0	$17,195	$0	$0	$70,560	$115,294
Total	$239,999	$30,677	$30,677	$99,292	$30,677	$30,677	$99,583	$561,582

Table 5.
Overview of input costs of 1 km lane of self-healing road [118].

self-healing ones. Over the road's lifespan, self-healing roads demonstrate lower costs in contrast to roads constructed using traditional conservation techniques.

6. Conclusions

As environmental sustainability becomes more significant in transportation infrastructure, both public and private contractors are increasingly acknowledging the importance of integrating sustainable practices into the design, construction, operation, and maintenance of highways and airports, including pavements. Embracing sustainable products, processes, and technologies in these areas offers significant advantages, not only promoting environmental well-being but also providing cost-saving opportunities for both government agencies and the construction industry. This not only improves the economic feasibility of national infrastructure development but also aligns with the increasing demand for durable, long-lasting roads with minimal maintenance needs. By prioritizing sustainability, stakeholders can create infrastructure that is resilient, environmentally friendly, and economically efficient, benefiting both present and future generations.

A growing emphasis in pavement construction revolves around incorporating waste materials to utilize them efficiently without compromising pavement quality. To address future priorities, such as accommodating rising traffic loads, it is imperative to explore new materials, embrace emerging technologies, and introduce innovative products. Anticipated challenges in the coming decades involve effectively managing the behavior and performance of asphalt mixtures, requiring a proactive approach to research, development, and implementation in the field of pavement construction. Worldwide, there is a continuous exploration of innovative methods and applications to enhance pavement longevity. These include emulsified bituminous mix, foam bituminous mix, fiber-reinforced bituminous mix, composite pavements, and perpetual pavements. One notable avenue of exploration is the use of self-healing asphalt, which holds potential benefits, especially in high-traffic areas where noise reduction is essential. However, the economic feasibility of this approach requires thorough validation to ascertain its justification for widespread implementation.

Self-healing asphalt technology for pavements utilizes the inherent self-repair capacity of asphalt activated by heat, where bitumen fills existing cracks. By postponing the conventional replacement of significantly deteriorated road surfaces, this technology not only avoids the production of new asphalt mixtures but also reduces energy consumption associated with replacement operations and minimizes the environmental impact of waste disposal, among other factors. Consequently, the implementation of microwave-healing technology is expected to lead to a significant decrease in atmospheric emissions. While there is practical experience with mobile microwave devices applied to pavements, the widespread adoption of this technology requires further studies to assess its environmental impact as an alternative to the conventional maintenance strategy centered on replacing deteriorated asphalt layers.

Acknowledgements

The authors would like to thank College of Engineering, King Faisal University, Al-Hasa, Kingdom of Saudi Arabia for their support.

Conflict of interest

The authors declare no conflict of interest.

Author details

Muhammad Aniq Gul[1*] and Muhammad Mudassir Ahmad Alwi[2]

1 Department of Civil and Environmental Engineering, College of Engineering, King Faisal University, Al-Hasa, Kingdom of Saudi Arabia

2 Department of Mechanical Engineering, College of Engineering, King Faisal University, Al-Hasa, Kingdom of Saudi Arabia

*Address all correspondence to: aniqgul@gmail.com

References

[1] Van Dam TJ, Harvey JT, Muench ST, Smith KD, Snyder MB, Al-Qadi IL, et al. Towards Sustainable Pavement Systems: A Reference Document. FHWA-HIF-15-002. Washington, DC: Federal Highway Administration; 2015

[2] Harvey JT, Meijer J, Ozer H, Al-Qadi IL, Saboori A, Kendall A. Pavement Life-Cycle Assessment Framework. FHWA-HIF-16-014. Washington, DC: Federal Highway Administration; 2016

[3] Chan WC. Economic and environmental evaluations of life-cycle cost analysis practices. In: A Case Study of Michigan DOT Pavement Projects. Ann Arbor, USA: University of Michigan; 2007

[4] Segundo IR, Freitas E, Branco VTFC, Landi S, Costa MF, Carneiro JO. Review and analysis of advances in functionalized, smart, and multifunctional asphalt mixtures. Renewable and Sustainable Energy Reviews. 2021;**151**:111552. DOI: 10.1016/j.rser.2021.111552

[5] Liu Y, Su P, Li M, You Z, Zhao M. Review on evolution and evaluation of asphalt pavement structures and materials. Journal of Traffic and Transportation Engineering (English Edition). 2020;**7**:573-599. DOI: 10.1016/j.jtte.2020.05.003

[6] Han B, Wang Y, Dong S, Zhang L, Ding S. Smart concretes and structures: A review. Journal of Intelligent Material Systems and Structures. 2015;**26**(11):1303-1345. DOI: 10.1177/1045389X15586452

[7] Han B, Zhang L, Ou J. Smart and Multifunctional Concrete toward Sustainable Infrastructures. Heidelberg: Springer; 2017. p. 548. ISBN 978-981-10-4348-2

[8] Rocha Segundo, Freitas, Landi Jr, Costa, Carneiro. Smart, photocatalytic and self-cleaning asphalt mixtures: A literature review. Coatings. 2019;**9**:696. DOI: 10.3390/coatings9110696

[9] Kalnæs SE, Jelle BP. Phase change materials and products for building applications: A state-of-the-art review and future research opportunities. Energy and Buildings. 2015;**94**:150-176. DOI: 10.1016/j.enbuild.2015.02.023

[10] Peng C, Yu J, Zhao Z, Fu J, Zhao M, Wang W, et al. Preparation and properties of a layered double hydroxide deicing additive for asphalt mixture. Cold Regions Science and Technology. 2015;**110**:70-76. DOI: 10.1016/j.coldregions.2014.11.013

[11] Baheri FT, Poulikakos LD, Poulikakos D, Schutzius TM. Ice adhesion behavior of heterogeneous bituminous surfaces. Cold Regions Science and Technology. 2021;**192**:103405. DOI: 10.1016/j.coldregions.2021.103405

[12] Pan P, Wu S, Xiao F, Pang L, Xiao Y. Conductive asphalt concrete: A review on structure design, performance, and practical applications. Journal of Intelligent Material Systems and Structures. 2015;**26**:755-769. DOI: 10.1177/1045389X14530594

[13] Luo S, Yang X. Performance evaluation of high-elastic asphalt mixture containing deicing agent Mafilon. Construction and Building Materials. 2015;**94**:494-501. DOI: 10.1016/j.conbuildmat.2015.07.064

[14] García Á, Schlangen E, van de Ven M, Liu Q. Electrical conductivity of asphalt mortar containing conductive fibers and fillers. Construction and Building Materials. 2009;**23**:3175-3181. DOI: 10.1016/j.conbuildmat.2009.06.014

[15] Giuliani F, Merusi F, Polacco G, Filippi S, Paci M. Effectiveness of sodium chloride-based anti-icing filler in asphalt mixtures. Construction and Building Materials. 2012;**30**:174-179. DOI: 10.1016/j.conbuildmat.2011.12.036

[16] Ma T, Geng L, Ding X, Zhang D, Huang X. Experimental study of deicing asphalt mixture with anti-icing additives. Construction and Building Materials. 2016;**127**:653-662. DOI: 10.1016/j.conbuildmat.2016.10.018

[17] Gao J, Sha A, Wang Z, Tong Z, Liu Z. Utilization of steel slag as aggregate in asphalt mixtures for microwave deicing. Journal of Cleaner Production. 2017;**152**:429-442. DOI: 10.1016/j.jclepro.2017.03.113

[18] Sun Y, Wu S, Liu Q, Hu J, Yuan Y, Ye Q. Snow and ice melting properties of self-healing asphalt mixtures with induction heating and microwave heating. Applied Thermal Engineering. 2018;**129**:871-883. DOI: 10.1016/j.applthermaleng.2017.10.050

[19] Peng C, Zhang H, You Z, Xu F, Jiang G, Lv S, et al. Preparation and anti-icing properties of a superhydrophobic silicone coating on asphalt mixture. Construction and Building Materials. 2018;**189**:227-235. DOI: 10.1016/j.conbuildmat.2018.08.211

[20] Arabzadeh A, Ceylan H, Kim S, Gopalakrishnan K, Sassani A. Superhydrophobic coatings on asphalt concrete surfaces: Toward smart solutions for winter pavement maintenance. Transportation Research Record. 2016;**2551**:10-17. DOI: 10.3141/2551-02

[21] Nahvi A, Sadoughi MK, Arabzadeh A, Sassani A, Hu C, Ceylan H, et al. Multi-objective Bayesian optimization of super hydrophobic coatings on asphalt concrete surfaces. Journal of Computational Design and Engineering. Oct 2019;**6**(4):693-704. DOI: 10.1016/j.jcde.2018.11.005

[22] Rocha Segundo I, Ferreira C, Freitas EF, Carneiro JO, Fernandes F, Júnior SL, et al. Assessment of 583 photocatalytic, superhydrophobic and self-cleaning properties on hot mix asphalts coated with TiO_2 and/or ZnO aqueous solutions. Construction and Building Materials. 2018;**166**:500-509. DOI: 10.1016/j.conbuildmat.2018.01.106

[23] Han S, Yao T, Yang X. Preparation and anti-icing properties of a hydrophobic emulsified asphalt coating. Construction and Building Materials. 2019;**220**:214-227. DOI: 10.1016/j.conbuildmat.2019.06.021

[24] Peng C, Hu X, You Z, Xu F, Jiang G, Ouyang H, et al. Investigation of anti-icing, anti-skid, and water impermeability performances of an acrylic superhydrophobic coating on asphalt pavement. Construction and Building Materials. 2020;**264**:120702. DOI: 10.1016/j.conbuildmat.2020.120702

[25] Schraufnagel DE, Balmes JR, Cowl CT, De Matteis S, Jung S-H, Mortimer K, et al. Air pollution and noncommunicable diseases. Chest. 2019;**155**:417-426. DOI: 10.1016/j.chest.2018.10.041

[26] Fan W, Chan KY, Zhang C, Zhang K, Ning Z, Leung MKH. Solar photocatalytic asphalt for removal of vehicular NOx: A feasibility study. Applied Energy. 2018;**225**:535-541. DOI: 10.1016/j.apenergy.2018.04.134

[27] Jensen H, Pedersen PD. Real-life field studies of the NOx removing properties of photocatalytic surfaces in roskilde and copenhagen airport, Denmark. Journal of Photocatalysis. 2020;**2**:71-81. DOI: 10.2174/2665976x01999200811155905

[28] European Environment Agency Emissions from Road Traffic and Domestic Heating behind Breaches of EU Air Quality Standards across Europe. Available from: https://www.eea.europa.eu/highlights/emissions-from-road-traffic-and

[29] Environmental Protection Agency Smog, Soot, and Other Air Pollution from Transportation. Available from: https://www.epa.gov/transportation-air-pollution-and-climate-change/smog-soot-and-other-air-pollution-transportation

[30] Chouhan J, Chandrappa AK. A systematic review on photocatalytic concrete for pavement applications: An innovative solution to reduce air pollution. Innovative Infrastructure Solutions. 2023;**8**:90. DOI: 10.1007/s41062-023-01060-6

[31] Hassan MM, Dylla H, Asadi S, Mohammad LN, Cooper S. Laboratory evaluation of environmental performance of photocatalytic titanium dioxide warm-mix asphalt pavements. Journal of Materials in Civil Engineering. 2012;**24**:599-605. DOI: 10.1061/(ASCE)MT.1943-5533.0000408

[32] Hassan M, Mohammad LN, Asadi S, Dylla H, Cooper S. Sustainable photocatalytic asphalt pavements for mitigation of nitrogen oxide and sulfur dioxide vehicle emissions. Journal of Materials in Civil Engineering. 2013;**25**:365-371. DOI: 10.1061/(ASCE)MT.1943-613 5533.0000613

[33] Fernández-Pampillón J, Palacios M, Núñez L, Pujadas M, Sanchez B, Santiago JL, et al. NOx depolluting performance of photocatalytic materials in an urban area—Part I: Monitoring ambient impact. Atmospheric Environment. 2021;**251**:616. DOI: 10.1016/j.atmosenv.2021.118190

[34] Hu J, Asce SM, Yu XB, Asce M. Reflectance spectra of thermochromic asphalt binder: Characterization and optical mixing model. Journal of Materials in Civil Engineering. 2015;**28**:1-10. DOI: 10.1061/(ASCE)MT.1943-5533.0001387

[35] Hu J, Yu X. Performance evaluation of solar-responsive asphalt mixture with thermochromic materials and nano-TiO_2 scatterers. Construction and Building Materials. 2020;**247**:118605. DOI: 10.1016/j.conbuildmat.2020.118605

[36] Lima O Jr, Freitas E, Cardoso P, Segundo IR, Margalho É, Moreira L, et al. Mitigation of urban heat island effects by thermochromic asphalt pavement. Coatings. 2023;**13**:1-17. DOI: 10.3390/coatings13010035

[37] Chen Z, Zhang H, Duan H, Shi C. Improvement of thermal and optical responses of short-term aged thermochromic asphalt binder by warm-mix asphalt technology. Journal of Cleaner Production. 2021;**279**:123675. DOI: 10.1016/j.jclepro.2020.123675

[38] Anupam BR, Sahoo UC, Rath P. Phase change materials for pavement applications: A review. Construction and Building Materials. 2020;**247**:118553. DOI: 10.1016/j.conbuildmat.2020.118553

[39] Ren J, Ma B, Si W, Zhou X, Li C. Preparation and analysis of composite phase change material used in asphalt mixture by sol–gel method. Construction

and Building Materials. 2014;**71**:53-62. DOI: 10.1016/j.conbuildmat.2014.07.100

[40] Ma B, Wang SS, Li J. Study on application of PCM in asphalt mixture. Advances in Materials Research. 2010;**168-170**:2625-2630. DOI: 10.4028/www.scientific.net/AMR.168-170.2625

[41] Chen MZ, Hong J, Wu SP, Lu W, Xu GJ. Optimization of phase change materials used in asphalt pavement to prevent rutting. Advances in Materials Research. 2011;**219-220**:1375-1378. DOI: 10.4028/www.scientific.net/AMR.219-220.1375

[42] Xu S, García A, Su J, Liu Q, Tabaković A, Schlangen E. Self-healing asphalt review: From idea to practice. Advanced Materials Interfaces. 2018;**5**:1800536. DOI: 10.1002/admi.201800536

[43] Lima O Jr, Cardoso P, Rocha Segundo I, Freitas E, Costa MFM, Nascimento JHO, et al. Thermochromism applied to transportation engineering: Asphalt roads, color, and road markings. In: Proceedings of the 5th International Conference on Application of Optics and Photonics (AOP); Guimarães, Portugal; 18-22 July 2022. IOP Publishing; 2022

[44] Chang X, Zhang R, Xiao Y, Chen X, Zhang X, Liu G. Mapping of publications on asphalt pavement and bitumen materials: A bibliometric review. Construction and Building Materials. 2020;**234**:117370. DOI: 10.1016/j.conbuildmat.2019.117370

[45] Pérez-Acebo H, Linares-Unamunzaga A, Abejón R, Rojí E. Research trends in pavement management during the first years of the 21st century: A bibliometric analysis during the 2000-2013 period. Applied Sciences. 2018;**8**:1041. DOI: 10.3390/app8071041

[46] Singer MN, Hamouda MA, El-Hassan H, Hinge G. Permeable pavement systems for effective management of stormwater quantity and quality: A bibliometric analysis and highlights of recent advancements. Sustainability. 2022;**14**:13061. DOI: 10.3390/su142013061

[47] Ali L, Alnajjar F, Khan W, Serhani MA, Al Jassmi H. Bibliometric analysis and review of deep learning-based crack detection literature published between 2010 and 2022. Buildings. 2022;**12**:432. DOI: 10.3390/buildings12040432

[48] El Hakea AH, Fakhr MW. Recent computer vision applications for pavement distress and condition assessment. Automation in Construction. 2023;**146**:104664. DOI: 10.1016/j.autcon.2022.104664

[49] Abejón R. Self-healing asphalt: A systematic bibliometric analysis for identification of hot research topics during the 2003-2018 period. Materials (Basel). 2021;**14**:565. DOI: 10.3390/ma14030565

[50] Su JF et al. Interface stability behaviors of methanol-Melamine-formaldehyde shell microPCMs/epoxy matrix composites. Polymer Composites. 2011;**32**:810-820

[51] Tabatabaee N, Shafiee MH. Effect of organoclay modified binders on fatigue performance. In: 7th RILEM International Conference on Cracking in Pavements. Delft, The Netherlands: RILEM Bookseries; 2012

[52] Garcia A, Jelfs J, Austin CJ. Internal asphalt mixture rejuvenation using capsules. Construction and Building Materials. 2015;**101**:8

[53] Jonkers HM, Schlangen E. A two components bacteria-based self-healing

concrete. In: Alexander MG, Beushausen H-D, Dehn F, Moyo P, editors. Concrete Repair, Rehabilitation and Retrofitting II. London: Tailor & Francis; 2009. pp. 119-120

[54] Dry C. Procedure developed for self-repair of polymer matrix composite materials. Composite Structures. 1996;**35**(3):263-239

[55] Wu M, Johnnesson B, Geiker M. A review: Self-healing in cementitious materials and engineered cementitious composite as a self-healing material. Construction and Building Materials. 2012;**28**:571-583

[56] Boh B, Šumiga B. Microencapsulation technology and its applications in building construction materials. Materials and Geoenvironment. 2008;**55**:329-344

[57] Kishi T et al. Self-healing behaviour by cementitious recrystallization of cracked concrete incorporating expansive agent. In: 1st International Conference on Self-Healing Materials, Noordwijk aan Zee, The Netherlands. MDPI; 2007

[58] Jin H et al. Self-healing thermoset using encapsulated epoxy-amine healing chemistry. Polymer. 2012;**53**(2):581-587

[59] Wilson GO, Magnus H. Self-healing systems for industrial and marine protective coatings. Journal of Protective Coatings & Linings. 2011;**28**(8):18-21

[60] Kirkby EL et al. Embedded shape-memory alloy wires for improvements of selfhealing. Advanced Functional Materials. 2008;**18**:2253-2260

[61] Lumley R. Self-healing in aluminium alloys. In: 1st International Conference on Self-Healing Materials, Delft, The Netherlands. 2007

[62] García A et al. A simple model to define induction heating in asphalt mastic. Construction and Building Materials. 2012;**31**:38-46

[63] García A et al. Electrical conductivity of asphalt mortar containing conductive fibers and fillers. Construction and Building Materials. 2009;**21**(10):3175-3181

[64] García Á, Schlangen E, van de Ven M. Two ways of closing cracks on asphalt concrete pavements: Microcapsules and induction heating. Key Engineering Materials. 2010;**417-418**:573-576

[65] García A, Schlangen E, van de Ven M. Induction heating of mastic containing conductive fibers and fillers. Materials and Structures. 2011;**44**(2):499-508

[66] Su JF, Schlangen E. Synthesis and physicochemical properties of high compact microcapsules containing rejuvenator applied in asphalt. Chemical Engineering Journal. 2012;**198-199**:289-300

[67] Su JF, Qiu J, Schlangen E. Stability investigation of self-healing microcapsules containing rejuvenator for bitumen. Polymer Degradation and Stability. 2013;**98**(6):1205-1212

[68] Su JF et al. Investigation the possibility of a new approach of using microcapsules containing waste cooking oil; in-situ rejuvenation. Construction and Building Materials. 2015;**74**:83-92

[69] Su JF, Qiu J, Schlangen E, et al. Experimental investigation of self-healing behavior of bitumen/microcapsule composites by a modified beam on elastic foundation method. Materials and Structures. 2015;**48**:4067-4076. DOI: 10.1617/s11527-014-0466-5

[70] Tabaković A et al. The use of compartmented sodium-alginate fibres as healing agent delivery system for asphalt pavements. In: Proceedings of European Materials Research Society Fall Meeting 2015; Warsaw, Poland

[71] Fisher H. Self repairing materials—Dream or reality. Natural Science. 2010;**2**(8):873-901

[72] Kim YRL, Little DN, Lytton RL. Fatigue and healing characterization of asphalt mixes. ASCE Journal of Materials in Civil Engineering. 2003;**15**(1):75-83

[73] Kök BV, Yilmaz M, Erkus Y. Effects of graphite on mechanical properties of stone mastic asphalt pavement. Journal of Civil Engineering and Management. 2017;**23**(8):1013-1020. DOI: 10.3846/13923730.2017.1374302

[74] García Á. Self-healing of open cracks in asphalt mastic. Fuel. 2012;**93**:264-272. DOI: 10.1016/j.fuel.2011.09.009

[75] Sun D, Hu J, Zhu X. Size optimization and self-healing evaluation of microcapsules in asphalt binder. Colloid and Polymer Science. 2015;**293**:3505-3516. DOI: 10.1007/s00396-015-3721-6

[76] Abadeen AZU, Hussain A, Sathish Kumar V, Murali G, Vatin NI, Riaz H. Comprehensive self-healing evaluation of asphalt concrete containing encapsulated rejuvenator. Materials. 2022;**15**:3672. DOI: 10.3390/ma15103672

[77] Gibney A. Analysis of Permanent Deformation of Hot Rolled Asphalt in Civil Engineering. University College Dublin; 2002

[78] Gibney A. Prediction of rutting resistance of hot rolled asphalt. In: 3rd Eurasphalt & Eurobitume Congress. Vienna, Austria: Foundation Eurasphalt; 2004

[79] Anderson FA. Final report on the safety assessment of melamine/formaldehyde resin. Journal of the American College of Toxicology. 1995;**14**(5):373-385

[80] Vázquez E, Barra M. Reactivity and expansion of electric arc furnace slag in their application in construction. Materiales de Construcción. 2001;**51**(263-264):137-148. DOI: 10.3989/mc.2001.v51.i263-264.359

[81] Jin J, Miao Y, Zhao H, Chen J, Qing L, Mu R, et al. Study on the self-healing performance of microcapsules and microcapsule-containing asphalt. Sustainability. 2022;**14**:12231. DOI: 10.3390/su141912231

[82] Cen. EN 13043/AC:2004. Aggregates for Bituminous Mixtures and Surface Treatments for Roads, Airfields and Other Trafficked Areas

[83] Fernández A, Alonso MA, López-Moro FJ, Moro MC. Polished stone value test and its relationship with petrographic parameters (hardness contrast and modal composition) and surface micro-roughness in natural and artificial aggregates. Materiales de Construcción. 2013;**63**(311):377-391. DOI: 10.3989/mc.2013.mc.00912

[84] Asi IM. Evaluating skid resistance of different asphalt concrete mixes. Building and Environment. 2007;**42**(1):325-329. DOI: 10.1016/j.buildenv.2005.08.020

[85] Pasetto M, Baldo N. Mix design and performance analysis of asphalt concretes with electric arc furnace slag. Construction and Building Materials. 2011;**25**(8):3458-3468. DOI: 10.1016/j.conbuildmat.2011.03.037

[86] Bosisio RG, Spooner J, Granger J. Asphalt road maintenance with a mobile

microwave power unit. Journal of Microwave Power. 1974;**9**(4):381-386. DOI: 10.1080/00222739.1974.11688936

[87] Al-Ohaly AA, Terrel RL. Effect of Microwave Heating on Adhesion and Moisture Damage of Asphalt Mixtures. Washington, DC, USA: National Research Council, Transportation Research Board; 1988

[88] Benedetto A, Calvi A. A pilot study on microwave heating for production and recycling of road pavement materials. Construction and Building Materials. 2013;**44**:351-359. DOI: 10.1016/j.conbuildmat.2013.02.082

[89] Wang Z, Zhao P, Ai T, Yang G, Wang Q. Microwave absorbing characteristics of asphalt mixes with carbonyl iron powder. Progress in Electromagnetics Research M. 2011;**19**:197-208. DOI: 10.2528/PIERM11052312

[90] Wang Z, Wang S, Ai T, Zhao P. Laboratory preparation and microwave heating test of CIPs/asphalt binder. In: Pavement Materials, Structures, and Performance. Vol. 239. GSP; 2014. pp. 1-8. DOI: 10.1061/9780784413418.001

[91] Gallego J, Del Val MA, Contreras V, Páez A. Heating asphalt mixtures with microwaves to promote self-healing. Construction and Building Materials. 2013;**42**:1-4. DOI: 10.1016/j.conbuildmat.2012.12.007

[92] Statistics. Association of Metallurgical Producers and Processors. Duisburg-Rheinhausen, Germany; 2010

[93] Franesqui MA, Yepes J, García-González C. Top-down cracking self-healing of asphalt pavements with steel fíler from industrial waste applying microwaves. Construction and Building Materials. 2017;**149**:612-620. DOI: 10.1016/j.conbuildmat.2017.05.161

[94] García A, Bueno M, Norambuena-Contreras J, et al. Induction healing of dense asphalt concrete. Construction and Building Materials. 2013;**49**:1-7. DOI: 10.1016/j.conbuildmat.2013.07.105

[95] Contreras JN, García A. Self-healing of asphalt mixture by microwave and induction heating. Materials & Design. 2016;**106**:404-414. DOI: 10.1016/j.matdes.2016.05.095

[96] Wang H, Apostolidis P, Liu X, et al. Laboratory test and numerical simulation of microwave heat-ing properties of asphalt mixture. In: Loizos A, Al-Qadi I, Scarpas T, editors. Bearing Capacity of Roads, Railways and Airfields. London: CRC Press; 2017. DOI: 10.1201/9781315100333-30

[97] Zhu X, Cai Y, Zhong S, et al. Self-healing efficiency of ferrite-filled asphalt mixture after microwave irradiation. Construction and Building Materials. 2017;**141**:12-22. DOI: 10.1016/j.conbuildmat.2017.02.145

[98] Sun Y, Wu S, Liu Q, et al. Self-healing performance of asphalt mixtures through heating fibers or aggregate. Construction and Building Materials. 2017;**150**:673-680. DOI: 10.1016/j.conbuildmat.2017.06.007

[99] Ayar P, Moreno-Navarro F, Rubio-Gámez MC. The healing capability of asphalt pave-ments: A state of the art review. Journal of Cleaner Production. 2016;**113**:28-40. DOI: 10.1016/j.jclepro.2015.12.034

[100] Sun Y, Liu Q, Wu S, Shang F. Microwave heating of steel slag asphalt mixture. Key Engineering Materials. 2014;**599**:193-197. DOI: 10.4028/www.scientific.net/KEM.599.193

[101] Contreras JN, Serpell R, Valdés Vidal G, et al. Effect of fibers addition on the physical and mechanical properties of asphalt mixtures with crack-healing purposes by microwave radiation. Construction and Building Materials. 2016;**127**:369-382. DOI: 10.1016/j.conbuildmat.2016.10.005

[102] Liu Q, Schlangen E, Van De Ven M, et al. Healing of porous asphalt concrete via induction heating. Road Materials and Pavement Design. 2010;**11**(sup1):527-542. DOI: 10.1080/14680629.2010.9690345

[103] Zhao H, Zhong S, Zhu X, et al. High-efficiency heating characteristics of ferrite-filled asphalt-based composites under microwave irradiation. Journal of Materials in Civil Engineering. 2017;**29**(6):04017007. DOI: 10.1061/(ASCE)MT.1943-5533.0001845

[104] Menozzi A, Garcia A, Partl MN, et al. Induction healing of fatigue damage in asphalt test samples. Construction and Building Materials. 2015;**74**:162-168. DOI: 10.1016/j.conbuildmat.2014.10.034

[105] Huang M, Huang W. Laboratory investigation on fatigue performance of modified asphalt concretes considering healing. Construction and Building Materials. 2016;**113**:68-76. DOI: 10.1016/j.conbuildmat.2016.02.083

[106] Contreras JN, Concha JL. Self-healing of asphalt mixtures via microwave heating. In: ISAP 2016 Symposium, Jackson, United States. 2016. Available from: https://www.researchgate.net/publication/305847215_Selfhealing_of_asphalt_mixtures_via_microwave_heating [Accessed: November, 2022]

[107] Liu Q, Schlangen E, Van de Ven M, et al. Evaluation of the induction healing effect of porous asphalt concrete through four point bending fatigue test. Construction and Building Materials. 2012;**29**:403-409. DOI: 10.1016/j.conbuildmat.2011.10.058

[108] García A, Contreras JN, Partl MN. A parametric study on the influence of steel wool fibers in dense asphalt concrete. Materials and Structures. 2013;**47**(9):1559-1571. DOI: 10.1617/s11527-013-0135-0

[109] Wang Z, Dai Q, Porter D, et al. Investigation of microwave healing performance of electrically conductive carbon fiber modified asphalt mixture beams. Construction and Building Materials. 2016;**126**:1012-1019. DOI: 10.1016/j.conbuildmat.2016.09.039

[110] Liu Q, Schlangen E, Ven MVD. Induction healing of porous asphalt concrete beams on an elastic foundation. Journal of Materials in Civil Engineering. 2013;**25**(7):880-885

[111] Pasetto M, Baldo N. Fatigue performance of asphalt concretes with RAP aggregates and steel slags. RILEM. 2012;**2012**:719-727. DOI: 10.1007/978-94-007-4566-7_70

[112] van Bochove G. Self-Healing Asphalt—Extending the Service Life by Induction Heating of Asphalt. Rosmalen, Netherlands: R&D, Heijmans Infra B.V.; 2016. DOI: 10.14311/EE.2016.310

[113] Porous Asphalt Pavement. 2014. Available from: https://www.slideshare.net/ashifar/porous-asphalt-pavement

[114] NL Agency, editor. Self Healing Materials Concept and Application. 2nd ed. The Hague: NL Agency, Dutch Ministry of Economic Affairs, Agriculture and Innovation; 2011

[115] Schlangen E. Other materials, applications and future developments.

In: Rooij ME et al., editors. Self-Healing Phenomena in Cement-Based Materials, RILEM Series: State-of-the-Art Reports. Dordecht: Springer; 2013. pp. 241-256

[116] Butt AA, Brigisson B, Kringos N. Optimizing highway lifetime by improving the selfhealing capacity of asphalt. Procedia - Social and Behavioral Sciences, Transport Research Arena. 2012;**48**:2190-2200

[117] Kringos N, Schmets A, Scarpas A, Pauli T. Towards understanding of the self-healing capacity of asphalt mixtures. Heron. 2011;**56**(1/2):45-74

[118] Ana M, Michael H, Jacob F, Juan G, Arne G, Richard W, et al. Consequences of long-term infrastructure decisions—The case of self-healing roads and their CO2 emissions. Environmental Research Letters. 2019;**14**:114040

Chapter 6

Natural Asphalt: A Novel Bitumen Modifier and Asphalt Filler

Mohammad Soleiman-Beigi, Homa Kohzadi and Saeed Toolabi

Abstract

Natural asphalt is a highly valuable material with diverse uses that humans have been utilizing for centuries. Its beneficial qualities, environmental friendliness compared to synthetic alternatives, versatility, and cost-effectiveness contribute to its overall importance as a valuable natural material. It emphasizes the abundance and significance of Iran's natural asphalt, which is known to be one of the largest producers globally. The investigations of the characteristics and analysis methods provide valuable information for further research and utilization in various industries, particularly in road construction and maintenance. The chapter utilizes several analysis methods, including Fourier Transform Infrared (FT-IR), elemental analysis, X-ray fluorescence, thermogravimetric analysis (TGA), and scanning electron microscope (SEM) analysis, to investigate the characteristics of Iranian natural asphalt. These techniques help understand its composition, thermal stability, and properties relevant to its application as a filler and bitumen modifier in asphalt mixtures and it explains how it improves various properties of traditional bitumen, enhancing stability, durability, rutting resistance, and water resistance. This is due to its exceptional traits such as high viscosity, thermoplastic behavior, and robust adhesive properties, making it a valuable modifier and filler in the industry.

Keywords: bitumen, natural asphalt, modifier, filler, softening point

1. Introduction

Bitumen modifiers and fillers play crucial roles in enhancing the performance of bitumen, which is a key component in road construction and pavement materials. These additives improve the durability, stability, and flexibility of asphalt mixtures, making them resistant to various environmental factors, such as heavy traffic, temperature fluctuations, moisture, and aging. The economic significance of bitumen modifiers and fillers lies in their ability to extend the lifespan of roads and pavements. By enhancing the performance of bitumen, these additives reduce the frequency of maintenance and repair works, leading to significant cost savings over the life cycle of a road. When bitumen is properly modified, it reduces the occurrence of cracks, rutting, and other forms of distress, which are common issues in traditional asphalt pavements. These results in longer-lasting roads and decreased maintenance expenses. Moreover, bitumen modifiers and fillers can enhance the performance of bituminous mixtures, allowing for thinner road surfaces. This reduces the overall material required for construction, resulting in direct cost savings in terms of reduced quantities of bitumen, aggregates, and other raw materials. Additionally, these additives improve the workability of bitumen during the construction process. They enhance the mixing and compaction properties of asphalt mixtures, making them easier to handle and place. This leads to reduced construction time and labor costs. Furthermore, bitumen modifiers and fillers can help in the use of recycled materials. By properly modifying bitumen, it becomes possible to incorporate reclaimed asphalt pavement (RAP) and other recycled materials into the asphalt mixtures. These promote sustainable practices by reducing the consumption of virgin materials and decreasing the overall construction costs. In recent years, there has been increasing interest in utilizing natural asphalt as a modifier or filler in various industries [1–4].

One of the primary applications is in road construction. Natural asphalt helps to lower the overall cost of road construction by reducing the need for other costly additives. Natural asphalt is used as a filler and modifier in various applications due to its exceptional binding properties. When mixed with other materials, it helps improve strength, durability, and flexibility, making it ideal for road construction, roofing, and waterproofing projects. For example, Little et al., in 2018, summarized the impact of fillers on the rheological properties of asphalt binders. They describe how active fillers, such as hydrated lime, can improve the asphalt-aggregate bond and magnify the volume filler effect. The chapter further explains how polymer additives interact with asphalt binders and how relatively low additive amounts of polymer by weight of asphalt binder can result in an inverted matrix [5]. Another study conducted by Gupta et al. in 2021 explored the impact of incorporating waste glass powder as new asphalt filler. Their findings indicated that the inclusion of glass powder enhanced the stiffness and rutting resistance of asphalt mixtures, thereby encouraging sustainable practices in the construction sector [6]. Furthermore, a significant study by Rubio-Gamez et al. in 2022 investigated the effects of filler type and bitumen on the mechanical properties of asphalt mortar. The research concluded that the filler type primarily influences the stiffness and ductility of the asphalt material, while bitumen plays a more significant role in the fatigue life of asphalt mortar [7].

The current chapter presents a comprehensive overview of natural asphalt as a valuable material, from its introduction and historical discovery to its global sources and analytical methods. Understanding its unique characteristics and properties is crucial for optimizing its usage and exploring its potential applications. The criteria for the research include The criteria for this research on natural asphalt include introduction and historical discovery of natural asphalt, global sources of natural asphalt, analytical methods used to study its characteristics and properties, applications of natural asphalt in the bitumen industry, understanding the significance and historical development of natural asphalt, current research trends and future prospects, highlighting the sustainable nature, versatility, durability, and environmental benefits of natural asphalt. The chapter aims to deepen our understanding of the historical significance, sustainability, versatility, durability, environmental benefits, and cost-effectiveness. The chapter explores its unique characteristics and properties, emphasizing its preferred usage in the preparation of bitumen and asphalt mixtures and by shedding light on the current research trends and promising future prospects, serves as a detailed exploration of the intriguing world of natural asphalt. In the subsequent, we will discuss the types of bitumen and solid natural asphalt. Also, we will explore its history and sources, investigate the characteristics of Iranian natural asphalt, and finally, examine its use as both a filler and modifier.

2. Definition of bitumen

The word bitumen is taken from asphaltic and related to the language of the ancient Akkad civilization. Ancient Iranians called bitumen mummy. The term asphalt has different meanings in Europe and South America. In Europe, the word asphalt means a mixture of bitumen and gravel. In South America, the word asphalt has been known to mean bitumen [8, 9].

2.1 Types of bitumen

Bitumen is a petroleum-based material that has various types and grades. Bituminous materials or heavy petroleum materials are prepared by a reaction with underground water and bacteria at relatively shallow depths. In such environments, oxygenated underground water dissolves and oxidizes part of the oil in the washing process. The bacteria in the water also break down less-weight molecules. During the phenomenon, which is called bacterial erosion, hydrogen is eliminated, and as a result, hydrocarbons and heavy oil molecules remain [10].

Bitumen is a heavy hydrocarbon material divided into different groups based on their physical properties, solubility in carbon tetrachloride, and chemical composition. Also, based on its resources, bitumen is classified into petroleum (refined) bitumen and natural bitumen.

2.1.1 Petroleum bitumen

Petroleum bitumen, also known as petroleum asphalt or refinery bitumen, is derived from the refining of crude oil. Crude oil is a complex mixture of hydrocarbons obtained from underground oil reserves. During the refining process, various fractions such as gasoline, diesel, kerosene, and bitumen are separated based on their boiling points.

Petroleum bitumen is a byproduct of the distillation of crude oil and is obtained as a residue from the atmospheric distillation process or vacuum distillation. It is composed of a mixture of heavy hydrocarbons and other impurities like sulfur, nitrogen, and trace metals. To improve its quality and remove impurities, petroleum bitumen undergoes further processing, such as air blowing, in which air is passed through the bitumen under controlled conditions. Petroleum bitumen is the most commonly used type of bitumen worldwide due to its abundance, ease of extraction, and standardized properties. It is used extensively in road construction, where it is mixed with aggregates to form asphalt concrete.

Tar also is a bitumen's material but it is different in origin and resource with bitumen. Tar is a black, high-viscosity liquid that is produced through the destructive distillation of organic materials such as coal or wood. It is composed of various organic substances and is sensitive to temperature changes, becoming hard and brittle in low temperatures and soft and liquefied in high temperatures. Due to its sensitivity to temperature, it is not ideal for areas with frequent temperature changes [11–13].

2.1.2 Natural bitumen (asphalt)

Natural bitumen, also known as asphaltite or natural asphalt, is referred to in this chapter by the same name. It originates from crude oil, which undergoes a process of gradual migration to the surface layers. Over time, low-weight materials evaporate, and under the influence of factors like heat, pressure, and various processes like oxidation, sulfurization, polymerization, and condensation, bitumen is formed (**Figure 1**) [11–13].

The results in the classification of bitumen into various grades with different applications. Natural bitumen based on physical properties is classified as tree types, as shown in **Figure 2** [11–13].

Bitumen can reach higher horizons and, in some cases, to the earth's surface under the pressure of the layers from inside the seams and cracks until it hardens completely. Coal and natural bitumen are both naturally occurring substances, but they have different properties and origins. The main difference between coal and natural bitumen is their physical form and composition. Coal is a solid fossil fuel composed mainly of carbon, while natural bitumen is a viscous liquid or semi-solid that is primarily composed of hydrocarbons.

In 1981, according to Abraham's theory, Jacob and Wenner classified natural asphalt with a degree of concentration based on the parameters of reflection, fluorescence, solubility, and softening temperature. The results are presented in **Table 1** [14].

2.1.2.1 Lake natural asphalt

Liquid natural bitumen, often referred to as crude bitumen or lake bitumen, is a highly viscous and sticky form of bitumen in its natural state. It is a thick, black, and tar-like substance with a lower melting point compared to its solid counterpart. It is typically stored and transported at high temperatures, around 150°C, as it becomes liquid and flows more easily at elevated temperatures. Liquid natural bitumen is usually found trapped in sand or sandstone reservoirs and can be extracted through various methods, including oil sand mining or *in situ* extraction techniques.

While solid natural bitumen is commonly used in asphalt applications and is solid at room temperature, liquid natural bitumen requires additional processing to transform it into a usable form, such as the production of synthetic crude oil or upgrading into lighter petroleum products. The viscosity and physical characteristics of solid and liquid natural bitumen differ significantly due to their distinct molecular structures and states [11–13]. Trinidad bitumen lake is an example of liquid bitumen with a relatively high viscosity, which means it is thick and sticky at normal temperatures. Its viscosity is highly temperature-dependent, becomes more fluid with increasing temperature, and solidifies at lower temperatures.

Figure 1.
Natura asphalt: Lump (a) lake (b).

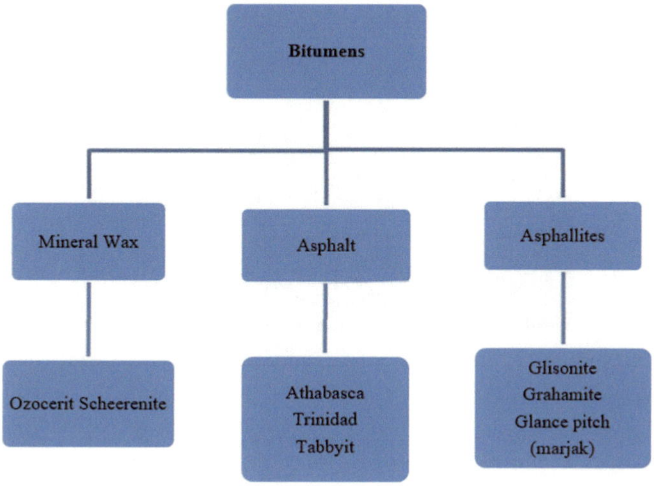

Figure 2.
Classification of natural bitumens (asphalts).

Natural asphalt	Subgroup of natural asphalt	Reflectance[a]	Fluorescence [b]	Solubility[c]	Softening Temperature [d]
Mineral Wax	Ozokerite	<0.02	9–50	insoluble	30–90
	Wurtzilite	<0.1	0.1–0.2	insoluble	no flow
	Albertite	0.1–0.7	<0.10	insoluble	no flow
Pyrobitumens	Impsonite Epi-impsonite	0.7–2.0	<0.02	insoluble	no flow
	meso-Impsonite	2.0–3.5	<0.01	insoluble	no flow
	kata impsonite	> 3.5	<0.01	insoluble	no flow
Asphalts	Asphalts	—	0.4–4.0	soluble	<104
	Gilsonite	0.07–0.11	0.05–0.40	soluble	104–164
Asphaltites	Glance pitch	0.11–0.30	0.05–0.20	soluble	104–164
	Grahamite	0.30–0.70	<0.05	soluble or insoluble	104–287

Notes:[a]Mean random reflectance oil immersion may be calculated from values obtained using water immersion.
[b]Fluorescence intensity at 546 nm where a masked uranyl glass standard (10 um diameter, Wild Leitz Co.) equals 1.00 intensity units.
[c]Observed solubility in immersion oil (or when cleaning specimen with petroleum ether).
[d]Observed hot-stage softening temperature (degree C).

Table 1.
Classification of natural asphalt based on reflectance, fluorescence, solubility, and softening temperature.

2.1.2.2 Solid natural asphalt

Solid natural bitumens, also known as asphaltites, are divided into three groups: gilsonite (uintaite), grahamite, and glanspit, which is one of the best and most famous groups of bitumen. It is very brittle and soft and turns into powder easily [15].

Solid natural asphalt is a solid mineral substance with scaly schists with shell fractures similar to volcanic glass or obsidian, which is black during extraction but turns brown after crushing and turning into powder (**Figure 3**). The mineral has a hydrogen content of about 9% and contains polar molecules whose ability to spread makes solid

Figure 3.
Solid natural asphalt veins.

natural asphalt soft at a temperature of more than 177°C. There are several analysis methods to determine the type or quality of natural asphalt [15].

2.1.2.3 Gilsonite

Gilsonite or uintaite is a brand name for a type of asphaltites that is black and shiny, which is found in a limited number of countries, such as Iran. Gilsonite is a naturally occurring, sticky hydrocarbon and natural asphalt, similar to hard petroleum asphalt, often referred to as natural asphalt. Gilsonite can dissolve in oily solvents as well as petroleum asphalt. Due to the compatibility of gilsonite, it is generally used to harden soft petroleum products. In bulk, gilsonite is a shiny black substance similar to the mineral obsidian. Gilsonite is fragile and can be crushed easily, and it turns into a brown powder [16]. The chemical analysis of gilsonite shows that in terms of elemental composition, it is composed of approximately 89.35% carbon, 191.9% hydrogen, 26.9% nitrogen, 27.1% sulfur, and 1.98% oxygen. **Figure 3** shows natural asphalt veins [17].

2.1.2.4 Pitch

Glance pitch, which is also called Manjak, has properties between gasolineite and grahamite, which means that it is heavier and has a higher melting point than gasoline, but the same properties of glance pitch are less than grahamite, and it has been found in Argentina, Cuba, America, Mexico, and Russia. Glance pitch is used in glazing and polishing as a kind of varnish, thermal plastic, coloring, and purification [18–20].

Petroleum pitch is a byproduct of the petroleum refining process and is primarily composed of aromatic hydrocarbons, including compounds like benzene, toluene, and xylene. It is obtained through the distillation and cracking of crude oil. The composition of petroleum pitch can vary based on the source of the crude oil and the refining methods used. It also contains impurities such as sulfur, nitrogen, and trace metals [18–20].

2.1.2.5 Grahamite

Grahamite was first discovered in a state of America and was later found in parts of Cuba, Trinidad, Colorado, and Oklahoma. Grahamite may have been obtained as a result of gasoline metamorphism. Graham is blacker than gasoline and has a shell fracture surface and many fractures. Grahamites are used as cement mortar, covering roofs and floors of buildings [21]. Grahamite has a Softening temperature of 104–287°C.

3. Chemical components of natural asphalt

Although determining the structure of natural asphalt is difficult, data analyses suggest natural asphalt comprises 70–80% of carbon and 15% of hydrogen, along with traces of nitrogen, sulfur, oxygen, and metals. Hydrocarbons can consist of polycyclic aromatic structures with different numbers of aromatic rings. Saturated hydrocarbon chains with different chain lengths and substitution patterns constitute the core of all these structures. Natural asphalt is generally characterized by aromatic fused rings, i.e., poly-nuclear aromatic, aliphatic side chains, and functional groups containing polar heteroatoms, e.g., amide, hydroxy, sulfide and sulfoxide, phenols, pyrroles, furanes, and pyridines (**Figure 4**) [22–25].

The different compounds of petroleum bitumen are not well known so far, but these compounds, which are separated from bitumen by means of different solvents, are divided into four general categories: saturated compounds, aromatic compounds, resins, and asphaltenes. Saturated compounds have saturated hydrocarbons, and aromatic compounds have simple rings. Resins and asphaltenes have similar chemical properties, and in the meantime, resins become asphaltene monomers, and asphaltene monomers together form asphaltene clusters [26, 27].

3.1 Saturated compounds

Chemically saturated compounds have linear, branched, and even cyclic hydrocarbon chains without any double bond or aromatic structure. Saturated compounds are mainly composed of carbon and hydrogen and have an H/C ratio of 1.8. Saturated compounds make up 5–20% of asphalt structure. From the physical point of view, tares are white to pale yellow viscous oils, and their molecular weight is around 800 g/mol [28].

3.2 Aromatic compounds

Aromatic compounds have unsaturated chemical structures and aromatic nuclei or rings. Aromatics account for 40–65% of asphalt structure. The molecular weight of these compounds is between 300 and 2000 g/mol, and the H:C ratio is equal to 1:5. Aromatics are created by connecting aromatic rings with non-polar and saturated carbon chains and act as solvents for heavy compounds in bitumen [29].

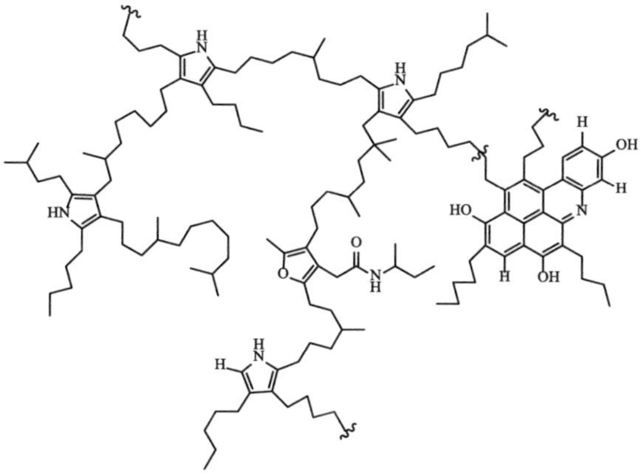

Figure 4.
A part of natural asphalt molecular structure.

3.3 Resins

Resins are compounds soluble in n-heptane and have a chemical structure similar to asphaltenes. Resins have a lower molecular weight than asphaltenes, and their H:C ratio is 4:1, higher than asphaltenes. Resins also contain oxygen and nitrogen compounds. Resins are dark brown compounds, solid to semi-solid, polar, and very viscous. Due to the structural similarity of resins to asphaltenes, they can also act as a dispersing agent for asphaltenes [30].

3.4 Asphaltenes

Asphaltenes are compounds made of aromatic rings welded together. Aliphatic bridges connect these planar polycyclic structures. With the advancement of chemical structure identification methods, serious differences have arisen regarding the molecular weight of asphaltenes. In the past, the molecular weight of asphaltenes was estimated in the range of 1000 to 10,000 g/mol, but nowadays, its value has decreased to 1000 to 2000 g/mol [30]. An example of saturated compounds (a), aromatic compounds(b), resins (c) and asphaltenes in structures of asphalt are shown in **Figure 5**.

4. History and discovery of natural asphalt

Natural asphalt has been utilized throughout history for various purposes. It was used by ancient civilizations such as the Sumerians, Persians, Egyptians, and Romans for waterproofing structures like roofs and walls. It was also mixed with aggregates to create durable roads, as seen in the road networks of the Romans, Incas, and Persians. Natural asphalt served as a sealant for ships, canoes, and containers, preventing leaks, and preserving contents. It was used in mummification in ancient Egypt to protect the body from decomposition. Additionally, bitumen was used in medicine, cosmetics,

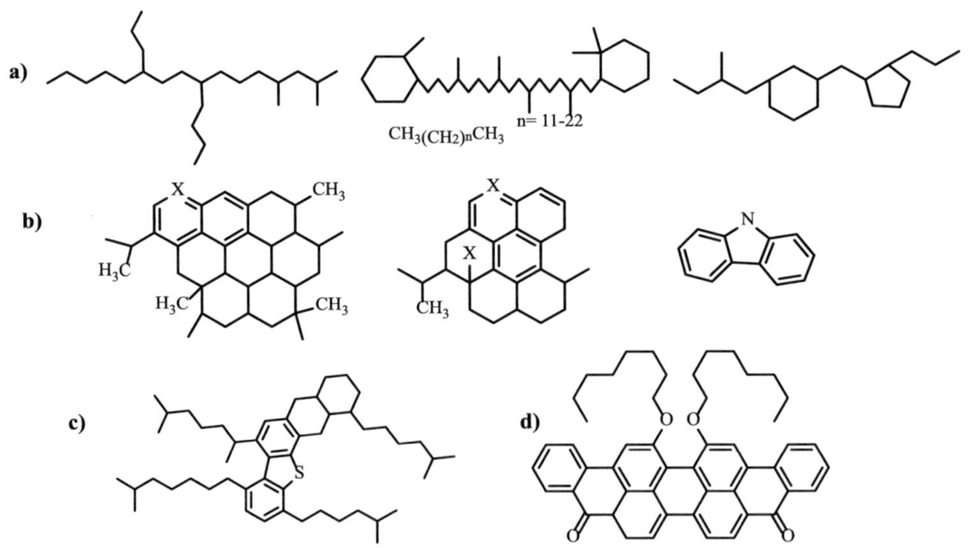

Figure 5.
Saturated compounds (a), aromatic compounds (b), resins (c) asphaltenes (d).

and artistic applications, acting as an ointment, base for cosmetics and perfumes, and a binder for pigments in paint. Natural bitumen's durability and adhesive properties made it a versatile material for ancient societies [31–36]. The natural leakage of ancient bitumen has been observed in various regions of the world specifically the Mesopotamian region (modern-day Iraq and Iran), is known for its natural seeps of bitumen (**Figure 6**). These regions were the first areas where natural bitumen leaks were discovered. Later, other new areas were discovered and expanded [37].

5. Natural asphalt in the world

Most reserves of natural bitumen mines are located in the United States, Canada, Iran, Iraq, Russia, Venezuela, China, Australia, Mexico, Albania, and the Philippines, respectively. Canada, Venezuela, and Russia have the most natural bitumen reserves. However, significant reserves are also concentrated in the United States, Mexico, Kuwait and Indonesia [33]. In the United States of America, the largest gilsonite reserves have been found in Utah and Colorado, estimated to have 45 million tons [38].

In Canada, technological advances have led to the recovery of bitumen in the remote areas of Northeastern Alberta [39].

There are many reserves and resources of heavy crude oil and natural bitumen, especially in Eastern Europe (Russia) and Siberia, where there are at least 700 billion barrels. Unlike the rest of the earth, most of the Russian Federation's reserves and resources are Mesozoic, Paleozoic, and Proterozoic. One hundred billion barrels of natural bitumen resources are located in China and Mongolia [40, 41].

Nigeria is a country in West Africa with many natural bitumen reserves. The Nigerian bitumen belt is found in the southwest of Nigeria and is located in the coastal areas of the eastern Dahomey basin, one of the prominent cities in Agababu, where natural bitumen was discovered for the first time in 1900. While the commercial exploitation of its materials has not yet begun, scientific research on it is steadily increasing [42].

Iran is one of the world's largest producers and exporters of solid natural bitumen. Iran holds significant reserves of natural bitumen, mainly found in the southwest region of the country, particularly in the provinces of Kermanshah and Ilam, western Iran. In Kermanshah, there are several natural asphalt sources scattered throughout

Figure 6.
The map of ancient natural asphalt and its applications.

Figure 7.
Iran's natural bitumen mining.

the province. Overall, the Kermanshah and Ilam provinces in Iran are rich in natural asphalt sources. These deposits contribute to the local economy and are utilized in various industries. The natural asphalt from these regions is of high quality and has gained recognition both domestically and internationally.

The natural bitumen deposits in Iran are classified into two types: lake bitumen and rock bitumen (solid natural asphalt). Rock bitumen or solid natural asphalt, on the other hand, is a major source of Iran's natural bitumen and is found in sedimentary carbonate rocks in various parts of Iran. These deposits are usually mined underground, and the bitumen is extracted using conventional mining techniques. The rock bitumen typically contains impurities such as sand, limestone, and shale, which are separated during the refining process. Lakes bitumen contains highly pure bitumen and is a minor source of Iran's natural bitumen production.

Iran has invested significantly in the extraction, processing, and marketing of natural bitumen. Iran's natural bitumen is known for its quality and diverse range of characteristics. **Figure 7** shows Iran's natural asphalt mine.

6. Characterization of natural asphalt

Various physical parameters such as ash content, carbon, hydrogen, nitrogen, and sulfur content, solubility, softening point, specific gravity and humidity are crucial in determining the quality of natural bitumen for different purposes [43]. The ash content in natural bitumen consists of inorganic materials and waste materials like clays, sulfides, sulfates, carbonams, etc. These materials remain as ash after combustion and are divided into inherent ash and secondary ash. The main elements in ash are aluminum, calcium, iron, sulfur, silicon, potassium, magnesium, sodium, phosphorus, and titanium [44].

Sulfur in natural bitumen, like coal, can exist in organic and inorganic forms. Inorganic sulfur includes sulfide minerals like pyrite, asphaltite, galena, and arsenopyrite, as well as sulfate minerals like barite, geysers, dendrite, and some iron sulfates. Pyrite sulfur is typically the dominant mineral sulfur in natural bitumen, existing as crystals among the bitumen structure. Organic sulfur in natural bitumen is covalently attached to its complex structure and is more difficult to remove compared to pyrite or mineral sulfur. It can exist as aromatic or aliphatic thiols, sulfides, disulfides, and hetero- and cyclic compounds [44].

Natural bitumen contains two types of moisture: outward moisture, which may be added during the extraction process, and analytical moisture, which is the intermolecular moisture present in natural asphalt [44]. Another important parameter is solubility. Solubility of natural bitumen depends on its large amount of asphaltene. It is soluble in aromatic hydrocarbons such as benzene, toluene, carbon disulfide, chloroform, and solvents containing chlorine. Natural bitumen has low solubility in alcohols and ketones. Solubility percentage in carbon disulfide (CS_2) can be used to determine the type and quality of mineral compounds present in natural bitumen. Carbon disulfide, trichloroethylene, and carbon tetrachloride are three commonly used solvents to assess the solubility percentage and quality of natural bitumen [44].

The softening point and specific gravity are important properties used to characterize and classify natural asphalt. The softening point of natural asphalt refers to the temperature at which it starts to soften and flow under the application of heat. It is a measure of the asphalt's resistance to deformation at elevated temperatures. The softening point is essential in determining the appropriate application temperature range for natural asphalt. It helps assess the material's ability to withstand temperature variations during storage, transportation, and use. Different grades of natural asphalt have different softening points, enabling them to be used in a wide range of applications, from road construction to roofing.

Specific gravity is the ratio of the density of a substance to the density of a reference substance, typically water. The specific gravity of natural asphalt provides information about its density and helps assess its quality. It is an important parameter in determining the asphalt's suitability for specific applications. As specific gravity is influenced by the composition and impurities in the asphalt, it can vary depending on the source and processing techniques. A high specific gravity indicates that the asphalt is denser, which makes it more durable and resistant to deformation. Low-specific gravity may indicate the presence of lighter hydrocarbons, which can lead to a less stable and weaker material. Overall, the softening point and specific gravity of natural asphalt are essential properties used to determine its performance and characteristics. These parameters assist in selecting the appropriate grade of asphalt for various applications, ensuring the material's optimal use and longevity [44].

Natural asphalt is chemically composed of complex organic compounds along with resins, oils, and waxes. The high hydrocarbon content in natural asphalt gives it its sticky and black characteristics. When heated, it undergoes molecular changes and can be used in various industrial processes. It softens when heated and hardens upon cooling. While generally chemically stable in its solid form, natural asphalt may experience oxidative degradation over time when exposed to air and sunlight, resulting in changes to its physical properties.

When studying natural asphalt, factors such as material characterization, cost, availability, challenges, and limitations must be considered. Material characterization involves studying its physical, chemical, and rheological properties. The cost of natural asphalt varies depending on extraction methods, refining processes, transportation, and market demand, which can be influenced by geological, economic, and political factors. Challenges and limitations include environmental impact and climate sensitivity due to extraction processes and weather conditions affecting its performance.

6.1 Characteristics of Iran's solid natural asphalt

Natural asphalt in Iran stands out from other types of asphalt due to its unique qualities. Composed primarily of hydrocarbon bitumen, it contains a high concentration of volatile components like naphthalene, toluene, and other aromatic compounds. The composition gives it advantageous properties, including a high melting point, low viscosity, and exceptional resistance to weathering. Renowned for its hardness and durability. Iran's solid natural asphalt is one of the hardest natural bitumen available. It can resist deformation under heavy traffic loads and withstand extreme weather conditions. In comparison, other types of asphalt often have lower hardness and require additives or modifiers to improve durability. Iran's natural asphalt also has a higher melting point, typically ranging from 180 to 245°C. This quality makes it suitable for applications in hot climates and areas with high-temperature variations. Other types of asphalt may have lower melting points, necessitating additional measures to prevent premature softening or melting. Additionally, Iran's natural asphalt exhibits excellent annular strength, which refers to its ability to adhere to aggregates or other materials used in asphalt mixtures. This property ensures good bonding and improved resistance to stripping or rutting. In contrast, other types of asphalt might require adhesion agents or chemical additives to achieve similar strength. Furthermore, Iran's natural asphalt possesses high chemical resistance due to its aromatic composition. It is less susceptible to chemical attack and degradation from fuel spillage, acidic substances, or saline water compared to other asphalts, making it a preferred choice for applications where exposure to corrosive substances is common. The solubility of Iran's natural asphalt may be slightly different than regular natural asphalt. Some studies have shown that Iran's natural asphalt has high solubility in organic solvents such as toluene, xylene, and carbon disulfide (**Figure 8**).

6.1.1 Mechanical analysis and structure study of natural asphalt of Iran

FT-IR (Fourier Transform Infrared) spectra of Iran's natural asphalt and petroleum bitumen have been studied. In the spectrum of Iran's natural asphalt, the broadband of 3200 cm^{-1} to 3600 cm^{-1} is related to NH and OH groups. The absorptions of 2920 cm^{-1} and 2850 cm^{-1} are, respectively, associated with the asymmetric and symmetric stretching vibration of the C-H bond of aliphatic compounds. The absorption of the 1600 cm^{-1} region is related to the stretching vibration of the C=C bond, which is another characteristic of asphaltene aromatic compounds. Absorptions in the regions of 1450 cm-1 and 1375 cm^{-1} are the vibrations of methyl groups (CH_3), which have been observed for both samples of Iran's natural bitumen and petroleum bitumen (**Figure 9**).

In another study, proton and carbon nuclear magnetic resonance (^1H and ^{13}C-NMR) samples of Iran's natural and petroleum bitumen were investigated. Natural

Figure 8.
Solid natural asphalt lumps (a) and a depot (b).

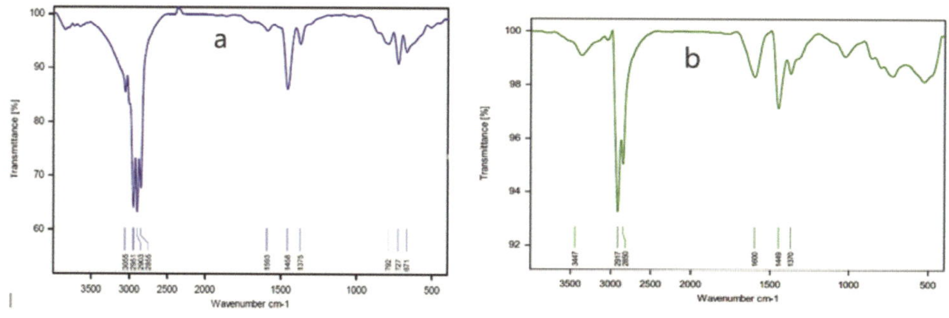

Figure 9.
FT-IR spectrum of petroleum bitumen (a) and solid natural asphalt (b).

bitumen is richer in terms of aliphatic (saturated) hydrogens and on the other hand, it has a small amount of aromatic hydrogens (**Figure 10**).

Also, elemental analysis and X-ray fluorescence of Iran's natural asphalt sample have been done. The results of these analyses are presented in **Tables 2** and **3**.

Thermogravimetric analysis (TGA) of Iran's natural asphalt shows that in the temperature range of 0–160°C, the sample did not lose weight. But in the 160°C and later can be considered as the beginning of the first step of decomposition of organic substances, which can be seen at 160–400°C, 10% of the organic substances in the sample have been decomposed. This step is called the first decomposition step. The graph below shows that the highest percentage of weight loss is assigned to the second step of decomposition, which is in the temperature range of 400–490°C, where 44% of the organic compound in the sample is decomposed. The third step of decomposition, in which a minimal amount of Iran's sample is decomposed, belongs to the temperature range of 490–810°C, during which another 7% of the compounds in the sample are decomposed. Finally, in a general conclusion, 61% of the organic substances in the sample have been destroyed in the temperature range of 0 to 810°C, and it is noteworthy that 39% of the organic compounds in the sample remain. The amount includes organic compounds with strong chemical bonds, as well as a percentage of ash. **Figure 11** shows the TGA analysis of Iran's natural asphalt. According

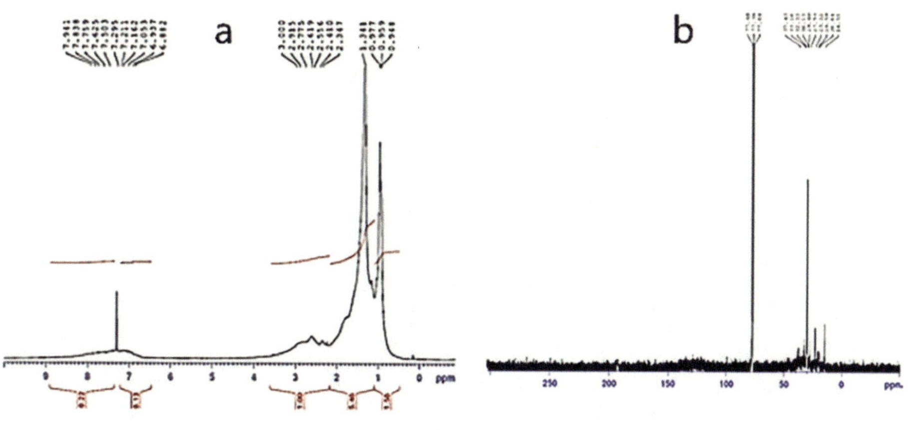

Figure 10.
^1H NMR and ^{13}C NMR spectrums of Iran's natural asphalt (400 MHz, CDCl3).

Element	C	H	S	N
%	75	12.5	2.5	2.3

Table 2.
The results of elemental analysis of Iran's natural asphalt.

Element	C	H	S	N	CaO	TiO_2	MnO	Al_2O_3	Fe_3O_4	K_2O	Na_2O	P_2O_5	MgO	SO_3	SiO_2
%	81.08	11.31	0.63	—	0.41	0.014	0.001	0.082	0.102	0.007	0.005	0.008	0.011	4.467	0.774

Table 3.
The results of X-ray fluorescence analysis of Iran's natural asphalt.

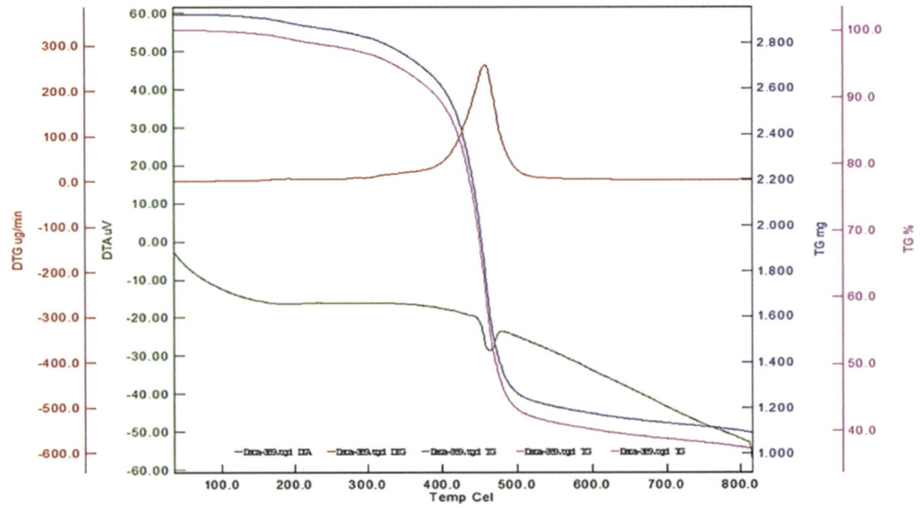

Figure 11.
TGA analysis of Iran's natural asphalt.

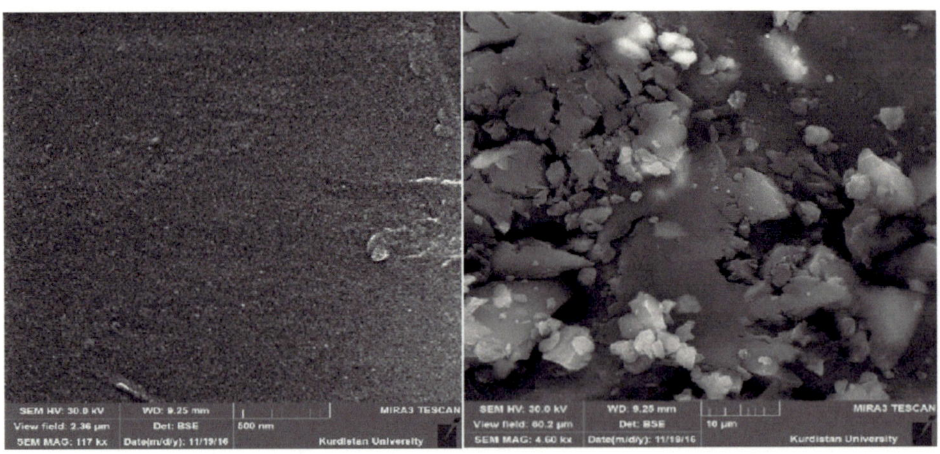

Figure 12.
SEM analysis of Iran's natural asphalt.

to the TGA analysis results, Iran's asphalt has good thermal stability and is suitable for use as a filler and bitumen modifier.

Iran's natural asphalt sample has been investigated using the scanning electron microscope (SEM) analysis method. The scanning electron microscope image of Iran's natural asphalt has shown the existence of multifaceted particles with smooth and flat surfaces (**Figure 12**) [44].

7. Applications of natural asphalt

Natural asphalt has been reported to have many valuable uses in various industries. Due to the fact that natural asphalt is a natural and non-toxic material, its use is increasing. The main applications of natural asphalt include the production of asphalt, paints and postings, isolation, preparation of emulsion bitumen, coke production, application in drilling mud, natural asphalt consumption as fuel, usage of natural asphalt in casting industries and usage of natural asphalt in the chemical industry. Most of the research carried out regarding the use of gilsonite in bitumen and asphalt was done by the American gilsonite company, which we will explain and mention in the following.

8. Petroleum bitumen modification

8.1 Natural asphalt (gilsonite), as a petroleum bitumen modifier

In the late 1970s, European contractors showed great interest in using bitumen modifiers because they believed in reducing the overall cost of using these modified bitumen compared to the high prices of raw materials. In the mid-1980s, more bitumen modifiers were investigated and known, attracting the special attention of American contractors. The use of modifiers gradually spread to different parts of the world [45]. Natural asphalt (gilsonite), can be used as a modifier in different forms, for example, gilsonite powder, granules, flakes, modified gilsonite, and blended gilsonite. Blended gilsonite is a mixture of natural asphalt and other additives or modifiers specifically formulated to achieve desired performance when added to petroleum bitumen.

The history of using bituminous modifiers dates back to 1943, when they were used for the first time. Natural asphalt can be used as a modifier in asphalt mixtures to enhance their performance characteristics. It is a performance-enhancing agent for asphalt mixes, achieving higher performance grades and easily blending into the asphalt without the need for high-shear milling. It also provides higher stability, reduced deformation, reduced temperature susceptibility, and increased resistance to water stripping compared to non-modified asphalt. Gilsonite can also be used to make pavement sealers with superior appearance and weathering. Its use in asphalt mix allows for a highly solid material without the need for additional balancing materials. Gilsonite can replace styrene-butadiene-styrene(SBS) polymers in asphalt modifiers, providing higher stability, lower shape changes, and increased resistance to water. It is commonly used due to its cost-effectiveness compared to other alternatives. When used as an asphalt binder modifier, gilsonite reduces susceptibility to high temperature and deformation performance issues and increases the use of recycled asphalt materials. Additionally, gilsonite-modified asphalt binder can produce a stronger road that is thinner compared to other pavements (**Figure 13**) [46].

On the other hand, when polymer-modified binders are used, they form large molecules that interconnect or cross-link with each other to create a matrix. This

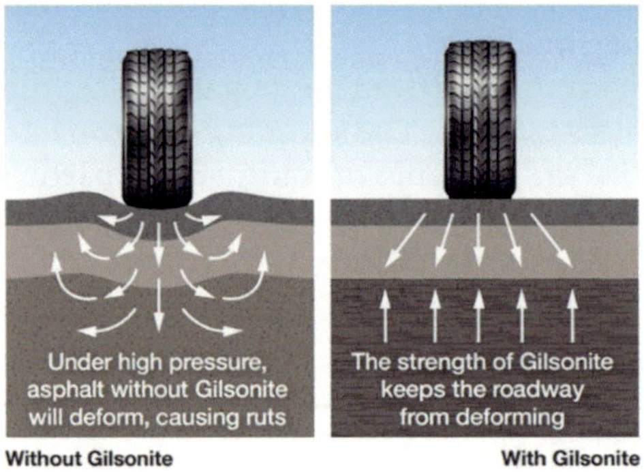

Figure 13.
Influence of gilsonite on the strength of asphalt.

matrix is not compatible with the recycled binder, meaning it cannot effectively bond with it. As a consequence, the pavement made with recycled asphalt and polymer-modified binders becomes more vulnerable to breaking down or deteriorating over time (**Figure 14**) [46].

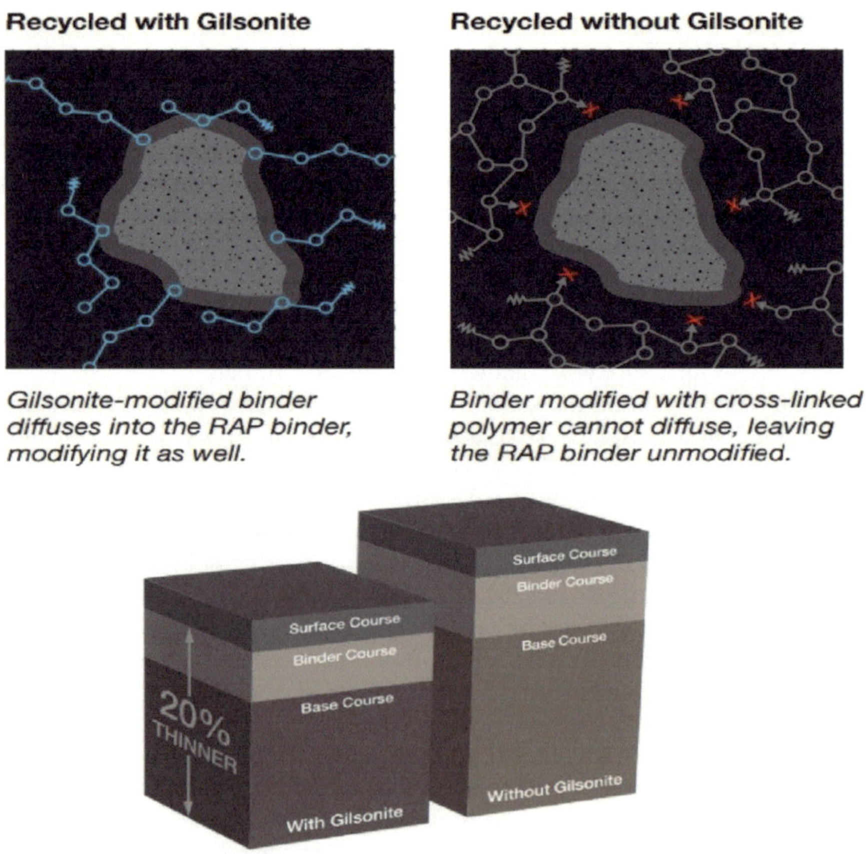

Figure 14.
Effect of gilsonite as a bitumen modifier.

So many studies have been performed in order to investigate the properties of modified bitumen using natural asphalt and (especially gilsonite). The below study shows the influence of different percentages of gilsonite in different weather conditions (temperature – 22 to +76) and also determines the performance grade (PG) of modified bitumen with gilsonite according to Superpio standards. The base bitumen used in the current project was PG 52–28 functional bitumen, which was modified in the presence of different percentages of gilsonite (0%, 3%, 6%, 9%, 12%) and was comprehensively investigated (**Table 4**) [47].

Considering that PG refers to the primary performance factors of asphalt, namely rutting, low temperature cracks, and aging cracks, it is highly regarded nowadays.

Some general properties and qualities of natural asphalt that make it suitable for modification include:

Natural asphalt has several properties that make it suitable for modification. It has a high softening point, enabling it to enhance stiffness and resistance at high temperatures. It exhibits thermoplastic behavior, softening when heated and hardening when cooled, making it a useful binding agent. Its elasticity and flexibility contribute to improved durability and crack resistance. Natural asphalt's chemical composition allows for effective modification and compatibility with other materials. It has strong adhesion and cohesion properties, enhancing strength and stability. It demonstrates high resistance to aging and weathering, ensuring long-term performance. Incorporating natural asphalt into asphalt mixtures increases flexibility and durability and enhances resistance to rutting and water infiltration. It can be modified through physical blending, chemical modification, oxidation, polymerization, and the use of additives [48].

8.2 Solid natural asphalt, as a petroleum bitumen filler

Some chemical specifications make it suitable as a filler material. These specifications are primarily related to its composition and physical properties. Solid natural asphalt primarily consists of hydrocarbons, including aromatic and aliphatic compounds. The presence of these hydrocarbons provides desirable adhesive and binding properties, making it an effective filler material. Also, natural asphalt has a high molecular weight, which contributes to its viscosity and solid-like behavior. This characteristic ensures that it remains in place and fills gaps effectively when used as filler. Another important factor is the chemical stability of natural asphalt. Natural asphalt is resistant to many chemicals, including acids, alkalis, and salts, which enhances its durability and resistance to degradation when used as a filler.

These chemical specifications and other physical and mechanical properties contribute to the effectiveness of natural asphalt as a filler material.

Physical specifications that make it suitable as a filler material. These physical properties contribute to its ability to fill gaps effectively and provide long-lasting performance. Here are the key physical specs of natural asphalt as filler:

Viscosity: Natural asphalt exhibits a suitable viscosity that allows it to flow and fill voids efficiently. Its viscosity can be adjusted by controlling the temperature during application, ensuring optimal workability as a filler.

Cohesion: Asphalt has inherent cohesive properties that enable it to bond together when used as an aggregate and maintain structural integrity. Additionally, it helps prevent the aggregate from separating or shifting over time.

Plasticity: Natural asphalt demonstrates plastic behavior, meaning it can be molded or shaped under pressure without breaking. The plasticity is advantageous for filling irregular cracks or voids, as it enables the filler to conform to the surrounding surfaces.

Gilsonite content	Flashpoint, °C	Mass loss, %	Viscosity @ 135°C	Grade Temp. at Which specified criterion is satisfied, °C						PG grade	
				DSR origin.	DSR RTFO	DSR PAV	BBR S	BBR m-value	DTT strain	High	Low
0	310	0.59	0.2168	52	52	16	−18	−18	—	52	−28
3	316	0.61	0.2555	58	58	19	−18	−18	—	58	−28
6	310	0.21	0.3400	58	58	22	−12	−12	—	58	−22
9	316	0.24	0.3842	64	64	25	−12	−12	—	64	−22
12	316	0.23	0.5208	70	70	25	−12	−12	—	70	−22

Table 4.
Summary of superpave binder test results.

Thermal Stability: Bitumen possesses excellent thermal stability, allowing it to withstand a wide range of temperatures without significant degradation.

Waterproofing Capability: Natural asphalt is impervious to water, making it an effective filler for sealing cracks and preventing moisture intrusion. It acts as a barrier, enhancing the durability and lifespan of structures.

Low Solubility: Bitumen exhibits low solubility in water and other common solvents. This characteristic contributes to the longevity of the filler material, as it resists dissolution or leaching out over time.

Durability: Asphalt is known for its exceptional durability, resistance to aging, and ability to withstand weathering effects such as UV radiation and oxidation. These properties ensure the filler maintains its performance and structural integrity over an extended period.

It is important to note that the physical specifications of natural asphalt may vary depending on the specific grade or source of the material. Additionally, the requirements for a filler material can vary based on the application and intended use [3].

Natural asphalt is used as a filler in various applications. As a filler, natural asphalt offers several benefits due to its unique properties. Here are some common uses of natural asphalt as filler:

- In the construction industry: Natural asphalt is often used as a filler in paving mixtures. It helps improve the durability, stability, and resistance of asphalt pavements. The addition of natural asphalt increases the viscosity and stiffness of the asphalt binder, leading to better performance in rutting resistance and fatigue cracking.

- In the production of adhesives and sealants: Natural asphalt is utilized as a filler in the formulation of adhesives and sealants. Its excellent adhesive properties and compatibility with various polymer systems make it an effective choice for enhancing the performance of these products. Natural asphalt helps improve the cohesion, tensile strength, and aging resistance of adhesives and sealants.

- In manufacturing inks and coatings: Natural asphalt is widely employed as a filler in the production of printing inks and coatings. It provides desirable properties like improved flow, pigment dispersion, and resistance to wear and tear. Natural asphalt acts as a rheological modifier, enhancing the handling and application characteristics of inks and coatings.

- In the rubber and tire industry: Natural asphalt is utilized as a filler and reinforcing agent in rubber compounds and tire manufacturing. Its incorporation improves the mechanical properties of rubber, such as tensile strength, abrasion resistance, and elasticity. Natural asphalt also enhances the processability and extrudability of rubber compounds [49, 50].

8.3 The effect of Gilsonite-modified asphalt on hot mix asphalt concert mixes

8.3.1 Study case: Texas

The District 12 laboratory conducted tests on different asphalt mix designs to assess the effects of gilsonite. In the current plan, gilsonite is added to the bitumen in percentages of 0%, 4%, 6%, and 8%. Modified bitumen is mixed with stone materials

Group	Average stability	Dry (psi)	Wet (psi)	Ratio wet/dry
Control 1	37	124.10	40.53	033
4% gilsonite	41	171.78	62.36	0.36
6% gilsonate	36	178.30	51.50	0.29
8% gilsonate	41	169.10	50.50	0.3
Design 2:				
Control 1	40	74.50	47.90	0.64
4% gilsonite	38	103.70	87..60	0.84
6% gilsonate	39	121.30	80.60	0.66
8% gilsonate	33	153.30	138.00	0.90

Table 5.
Results from mix design 1.

according to patterns 1 and 2. The stability, specific gravity, indirect elasticity and moisture sensitivity of the resulting asphalt mixture have been investigated (**Table 5**) [51]. The results of the laboratory tests showed that the gilsonite-modified asphalt mixes increased both dry and wet tensile strength. As there is not sufficient evidence to prove that gilsonite alone is effective as an anti-stripping agent (preventing the separation of asphalt and aggregate due to moisture), it is recommended to use it in conjunction with an anti-stripping agent to mitigate the risk of stripping and rutting.

Also, it was found that gilsonite did not have a significant impact on Hveem stability, which is a measure of the resistance to deformation under loading (**Figure 15**). To determine Hveem stability, a cylindrical specimen of the asphalt mixture is subjected to a vertical load in a controlled testing apparatus. The load is gradually applied, and the specimen's resistance to deformation is measured. During the test, the stability

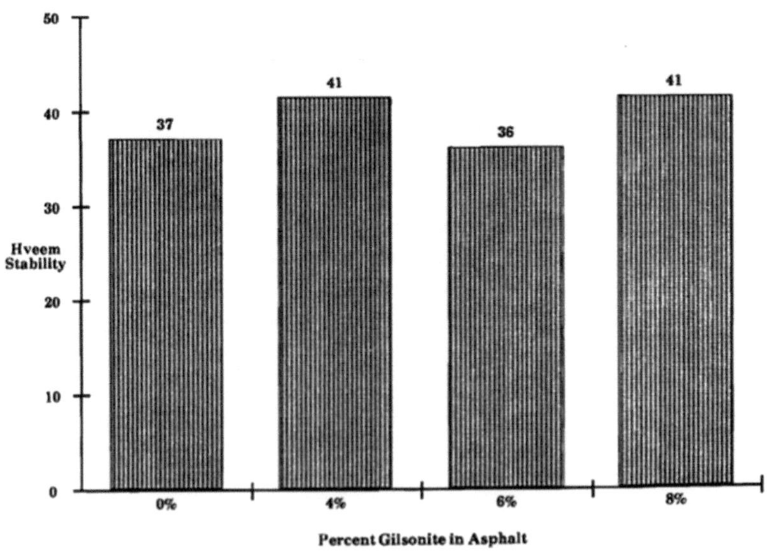

Figure 15.
Stability vs. percent gilsonite in asphalt mix design 1.

value represents the maximum load sustained by the specimen without causing excessive deformations or failure. A higher stability value indicates better resistance to deformation and is desirable for durable asphalt pavements. Hveem stability is influenced by various factors, including the type and amount of aggregate used, the proportion of asphalt binder, the level of compaction, and the presence of additives, such as gilsonite in this case. In the current project, it was found that the addition of gilsonite did not significantly increase the Hveem stability of the asphalt mixes. This suggests that gilsonite may not have a substantial impact on the resistance of the asphalt mixture to deformation under loading.

9. Comparing the use of asphalt as filler and modifier

As a clarification, natural asphalt is typically used as a filler rather than a modifier in various applications. It is more commonly utilized for its physical properties and benefits as a filler material rather than as a modifier to alter the characteristics of a material. However, suppose you are specifically interested in the drawbacks of using natural asphalt as a modifier. Natural asphalt may not have the desired properties to modify materials effectively. It may have limited abilities compared to synthetic modifiers in achieving specific performance enhancements. Natural asphalt can vary in composition and properties, making it challenging to achieve consistent results. Compatibility issues may arise with certain materials or chemical additives, potentially reducing effectiveness or causing negative interactions.

9.1 Methods of modification of petroleum bitumen with natural asphalt (gilsonite)

Mixing method refers to the process of combining different materials in an asphalt plant to produce asphalt mixture. One of the key components in asphalt mixture is bitumen and is the binding agent that holds the asphalt together. In an asphalt plant, the process of mixing bitumen with other materials (such as aggregate, filler, and additives) takes place in the asphalt mixer. Different mixing methods can be employed, depending on the type of asphalt plant and the desired properties of the asphalt mixture. One common mixing method used in asphalt plants is the batch mixing method. In this method, the different materials, including bitumen, are weighed and then added to the mixer. The mixer ensures that all the materials are thoroughly mixed together to form a consistent and uniform mixture. The process is repeated for each batch of asphalt mixture produced.

Another mixing method commonly used in asphalt plants is the continuous mixing method. The materials are continuously fed into the mixer, including the bitumen. The mixer has rotating paddles or blades that mix and blend the materials together as they pass through the mixer. This ensures a continuous flow of asphalt mixture output.

The choice of mixing method depends on various factors, such as the plant capacity, the type of asphalt mixture being produced, and the construction project requirements. Both batch and continuous mixing methods have their advantages and disadvantages. For example, batch mixing allows for more control and flexibility in batch size and mixture composition, while continuous mixing offers higher production rates. For example, in the following method, 200 mesh micronized powder is added to the flask of molten refinery bitumen at 170°C and stirred for 2 hours by a regular stirrer. The method is for percentages less than 12% (**Figure 16**) [52].

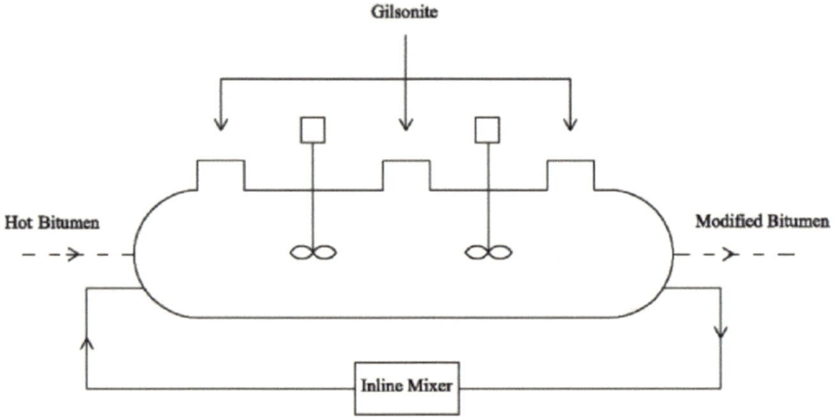

Figure 16.
Method of modification of petroleum bitumen by gilsonite.

9.2 Preparation of modified asphalt using natural asphalt powder (Gilsonite)

In this method, gilsonite is stored in a silo that is part of the plant, and stone materials and natural asphalt powder (gilsonite) enter the mixer of the asphalt plant at the same time; therefore, the time of injection or spraying of natural asphalt powder (gilsonite) into the mixer is adjusted at the same time as the stone materials enter the mixer. The temperature of stone materials in the batch mixer should be between 160 and 174°C, so the temperature in that range should be adjusted and fixed (**Figure 17**) [52].

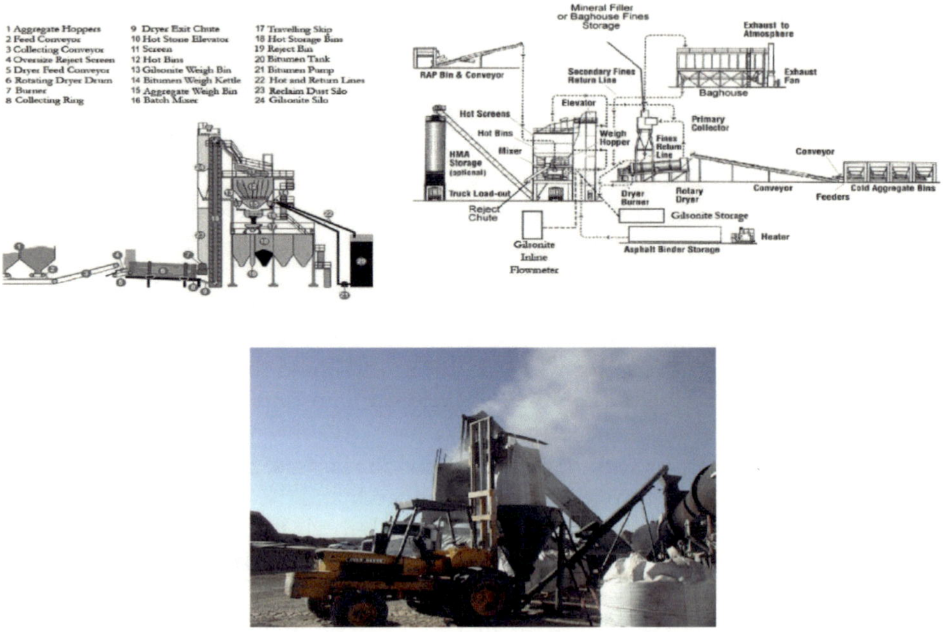

Figure 17.
Addition of gilsonite to bitumen.

10. Future of natural asphalts

Natural asphalt has been primarily used in various applications such as road construction, roofing materials, waterproofing, and industrial compounds. However, it is important to note that industries and technologies are constantly evolving, and the future of natural asphalt applications may involve new developments and trends. Here are a few potential directions for the future of natural asphalt applications:

1. Sustainable Infrastructure: With increasing global emphasis on sustainability, there may be a growing demand for eco-friendly alternatives in infrastructure development. Natural asphalt, a naturally occurring material, could gain popularity as a renewable and sustainable option for road pavements and other construction projects. Efforts to enhance the durability and performance of natural asphalt through research and innovation could further drive its adoption.

2. Advanced Manufacturing Techniques: Advancements in manufacturing techniques, such as nanotechnology and polymer modification, could offer new possibilities for natural asphalt applications. These techniques can help enhance the strength, flexibility, and overall performance of natural asphalt, making it suitable for a wider range of applications.

3. Green Roofing: As green building practices gain momentum, natural asphalt could find increased use in green or vegetated roofing systems. Natural asphalt can provide an effective waterproofing layer while complementing the overall sustainability goals of green roofs.

4. Alternative Energy Storage: Natural asphalt's thermal properties make it a potential candidate for energy storage applications. It can absorb and release heat effectively, presenting opportunities for using natural asphalt in thermal energy storage systems or solar heating applications.

It is worth noting that these projections are speculative, and the actual future of natural asphalt applications may depend on several factors, such as market demand, technological advancements, environmental regulations, and availability of alternative materials.

11. Conclusion

In conclusion, the chapter on introducing natural asphalt provides valuable information on its characteristics, sources, analysis methods, and applications. Natural asphalt, also known as gilsonite and asphaltite, is a unique geological material with diverse properties and applications. The chapter highlights the various sources of natural asphalt around the world, emphasizing the abundance and significance of Iran's natural asphalt. Iran is known as one of the largest producers of natural asphalt, with vast reserves and a long history of utilization in various industries. The characteristics of Iranian natural asphalt were investigated using FT-IR, elemental analysis, X-ray fluorescence, TGA, and SEM analysis. According to the elemental analysis and X-ray fluorescence of Iran's natural asphalt, it has a high percentage of carbon. Additionally, the TGA analysis results have shown that Iran's asphalt has good thermal stability and is suitable for use as a filler and bitumen modifier. Asphalt modifiers can enhance the performance of asphalt by improving its stiffness, durability, and resistance to aging. The addition of asphalt fillers can improve the volumetric properties of asphalt mixtures, such as stability and void content. The effectiveness can vary depending on factors like climate, traffic conditions, and mix design.

Various types of modifiers, such as polymers, crumb rubber, and natural fibers, have been studied and shown to have positive effects on asphalt performance. The use of certain modifiers, such as crumb rubber from scrap tires, raises environmental and sustainability concerns. The use of bitumen modifiers and asphalt fillers holds promise for improving the performance of asphalt mixtures. However, further research is required to determine the optimal materials and dosage for specific conditions. Considerations regarding environmental impact and sustainability are important in the selection of modifiers and fillers.

Natural asphalt has proven to be an effective additive in enhancing the properties of traditional bitumen, such as improving stability, durability, and rutting resistance in asphalt mixtures. It has also been utilized as an alternative to bitumen in road construction and maintenance, offering economic and environmental advantages. Natural bitumen has several characteristics that make it suitable for modifiers and fillers. Natural asphalt has a high viscosity, making it an effective modifier for materials like asphalt, rubber, and polymers. It exhibits thermoplastic behavior, softening when heated and hardening when cooled, enhancing flexibility and durability. Bitumen is highly resistant to water penetration, improving water resistance in materials. Additionally, it has excellent adhesive properties, enhancing bonding strength and improving material performance. However, there are a few disadvantages of using bitumen and asphalt as a modifier and filler. Over time, bitumen and asphalt can undergo aging and weathering, causing them to become brittle and prone to cracking. That can result in the deterioration of road surfaces and the need for frequent repairs. Also, bitumen and asphalt surfaces can contribute to increased noise levels and vibration, particularly in urban areas with high traffic volumes, and it have adverse effects on nearby residents and sensitive structures.

Overall, the chapter underscores the importance of natural asphalt as a valuable resource with unique characteristics and wide-ranging applications. Its abundance in Iran and the analysis methods discussed pave the way for further research and utilization of natural asphalt in various industries, particularly in the field of road construction and maintenance.

Author details

Mohammad Soleiman-Beigi[1*], Homa Kohzadi[1] and Saeed Toolabi[2]

1 Department of Chemistry, Faculty of Basic Sciences, Ilam University, Ilam, Iran

2 Department of Civil Engineering, Master of Engineering, Ilam University, Ilam, Iran

*Address all correspondence to: soleimanbeigi@yahoo.com; m.soleimanbeigi@ilam.ac.ir

References

[1] Luo J, Li Q, Ali MA, Alizadeh AA, Raise A, Alali AF, et al. Experimental investigation of the rheological behaviors of bitumen and stone matrix asphalt mixtures using waste material. Construction and Building Materials. 2023;**15**(365):130003. DOI: 10.1016/j.conbuildmat.2022.130003

[2] Wen Y, Wang Y, Zhao K, Sumalee A. The use of natural rubber latex as a renewable and sustainable modifier of asphalt binder. International Journal of Pavement Engineering. 2017;**18**:547-559. DOI: 10.1080/10298436.2015.1095913

[3] Wu W, Jiang W, Yuan D, Lu R, Shan J, Xiao J, et al. A review of asphalt-filler interaction: Mechanisms, evaluation methods, and influencing factors. Construction and Building Materials. 2021;**299**:124279. DOI: 10.1016/j.conbuildmat.2021.124279

[4] Liu N, Liu L, Li M, Sun L. Effects of zeolite on rheological properties of asphalt materials and asphalt-filler interaction ability. Construction and Building Materials. 2023;**382**:131300. DOI: 10.1016/j.conbuildmat.2021.124279

[5] Sun M, Wang J, Sun H, Hong B. Feasibility analysis of polyurethane-Prepolymer-modified bitumen used for fully reclaimed asphalt pavement (FRAP). Materials. 2023;**16**:5686. DOI: 10.3390/ma16165686

[6] Choudhary J, Kumar B, Gupta A. Utilization of waste glass powder and glass composite fillers in asphalt pavements. Advances in Civil Engineering. 2021;**13**:1-7. DOI: 10.1155/2021/3235223

[7] Tauste-Martínez R, Hidalgo AE, García-Travé G, Moreno-Navarro F, Rubio-Gámez MD. Influence of type of filler and bitumen on the mechanical performance of asphalt mortars. Materials. 2022;**15**:3307. DOI: 10.3390/ma15093307

[8] Asphalt Institute, The Asphalt Handbook MS-4, Asphalt Institute. 7th ed, Lexington, KY (Kentucky). 2007. ISBN-10: 193415427X 2007

[9] Read J, Whiteoak D. The Shell Bitumen Handbook. 5th ed. London: Thomas Telford Publishing; 2003. p. 2553

[10] Delano WH. Twenty Years' Practical Experience of Natural Asphalt and Mineral Bitumen. Whitefish, MT: Kissinger Publishing; 2007 (original 1893)

[11] Wen CS, Chilingarian GV, Yen TF. Properties and structure of bitumens. Bitumens, asphalts and tar sands. 1978;**7**:155. DOI: 10.1016/S0376-7361(08)70066-9

[12] Bitumen [Internet]. Wikipedia. 2023. Available from: https://en.wikipedia.org/wiki/Bitume

[13] Tar pit [Internet]. Wikipedia. 2023. Available from: https://en.wikipedia.org/wiki/Tar_pit

[14] Abraham H. Asphalts and Allied Substances, sixth edition, volume 1, Van Nostrand, Princeton. 1960. 370 p

[15] Erdman JG, Ramsey VG. Rates of oxidation of petroleum asphaltenes and other bitumens by alkaline permanganate. Geochimica et Cosmochimica Acta. 1961;**25**:175. DOI: 10.1016/0016-7037(61)90075-8

[16] Ameri M, Mansourian A, Ashani SS, Yadollahi G. Technical study on the Iran's Gilsonite as an additive for modification

of asphalt binders used in pavement construction. Construction and Building Materials. 2011;**25**:1379. DOI: 10.1016/j.conbuildmat.2010.09.005

[17] Kohzadi H, Soleiman-Beigi M. A recyclable heterogeneous nanocatalyst of copper-grafted natural asphalt sulfonate (NAS@ Cu): Characterization, synthesis and application in the Suzuki–Miyaura coupling reaction. New Journal of Chemistry. 2020;**44**:12134. DOI: 10.1039/D0NJ01883J

[18] Aureal T, Cross D, Gordon D. Palynology and petrography of solid Bitumens of the Uinta Basin, Utah. Brigham Young University Geology Studies. 1976;**22**:157-173

[19] Chockalingam K, Saravanan U, KrishnanJM.Characterizationofpetroleum pitch using steady shear experiments. International Journal of Engineering Science. 2010;**48**:1092-1109. DOI: 10.1016/j.ijengsci.2010.08.008

[20] Han SH, Park EJ. Characterization and application of petroleum pitch. Materials. 2014;**7**:816-826. DOI: 10.1039/C8RA04679D

[21] Jacob H. Classification, structure, genesis and practical importance of natural solid oil bitumen ("migrabitumen"). International Journal of coal geology. 1989;**11**:65. DOI: 10.1016/0166-5162(89)90113-4

[22] Kohzadi H, Soleiman-Beigi M. Immobilization of PdCl2 on a natural asphalt sulfonic acid network for C–N and C–O bonds formation. ChemistrySelect. 2022;**7**:e202200799. DOI: 10.1002/slct.202200799

[23] Kohzadi H, Soleiman-Beigi M. Progress on the natural asphalt applications as a new class of carbonious heterogeneous support; synthesis of Na [Pd-NAS] and study of its catalytic activity in the formation of carbon–carbon bonds. Molecular Diversity. 2022;**26**:1957. DOI: 10.1007/s11030-021-10306-3

[24] Kohzadi H, Soleiman-Beigi M. XPS and structural studies of Fe3O4-PTMS-NAS@ Cu as a novel magnetic natural asphalt base network and recoverable nanocatalyst for the synthesis of biaryl compounds. Scientific Reports. 2021;**30**(11):24508. DOI: 10.1038/s41598-021-04111

[25] Falah S, Soleiman-Beigi M, Kohzadi H. Potassium natural asphalt sulfonate (K-NAS): Synthesis and characterization as a new recyclable solid basic nanocatalyst and its application in the formation of carbon–carbon bonds. Applied Organometallic Chemistry. 2020;**34**:e5840. DOI: 10.1002/aoc.5840

[26] Whiteoak CD. The Shell Bitumen Industrial Handbook. Surrey: Shell Bitumen; 1995

[27] Mukhametshin RZ, Punanova SA. Composition of natural bitumens from the Ural-Volga region. Solid Fuel Chemistry. 2014;**48**:58-70. DOI: 10.3103/S0361521914010078

[28] Speight JG. The Chemistry and Technology of Petroleum. 4th. ed. Boca Raton: CRC Press; 2006. DOI: 10.1201/9781420008388

[29] Read J, Whiteoak D, Hunter RN. The Shell Bitumen Handbook. Fifth ed. London, UK: Thomas Telford; 2003

[30] Krchma LC, Gagle DW. A USA history of asphalt refined from crude oil and its distribution. In: Association of Asphalt Paving Technologists Proc. University of Minnesota. Vol. 43. 1974 Accession Number: 00127522

[31] Chen JS, Huang LS. Developing an aging model to evaluate engineering properties of asphalt paving binders. Materials and Structures. 2000;**33**:559-565. DOI: 10.1007/BF02480536

[32] Krishnan M, J, Rajagopal KR. Review of the uses and modeling of bitumen from ancient to modern times. Applied Mechanics Reviews. 2003;**56**:149-214. DOI: 10.1115/1.1529658

[33] Clark KA, Ikram S, Evershed RP. The significance of petroleum bitumen in ancient Egyptian mummies. Philosophical Transactions of the Royal Society A: Mathematical, Physical and Engineering Sciences. 2016;**374**:20160229. DOI: 10.1098/rsta.2016.0229

[34] Edwards DS, Vinall DR, Corrick AJ, McKirdy DM. Natural bitumen stranding on the ocean beaches of southern Australia: A historical and geospatial review. Transactions of the Royal Society of South Australia. 2016;**140**:152. DOI: 10.1080/03721426.2016.1203532

[35] Roberts JD. Performance of Cement-Modified Soils: A Follow-Up Report. Oklahoma City: Transportation Research Board. 1986. pp. 81-86. DOI: worldcat.org/isbn/0309041090

[36] Crawford AL. Gilsonite-its Discovery and the Early History of the Industry. The United States: Eighth Annual Field Conference; 1957

[37] Connan J. Use and trade of bitumen in antiquity and prehistory: Molecular archaeology reveals secrets of past civilizations. Philosophical transactions of the Royal Society of London. Series B: Biological Sciences. 1999;**354**:33-50. DOI: 10.1098/rstb.1999.0358

[38] Rahimi E, Shekarian Y, Farahani S, Asgari G, Nakini A. New approach in application of the AHP–fuzzy TOPSIS method in mineral potential mapping of the natural bitumen (Gilsonite): A case study from the Gilan-e-Gharb block, the Kermanshah, west of Iran. American Journal of Engineering and Applied Sciences. 2020;**13**:96. DOI: 10.3844/ajeassp.2020.96.110

[39] Ameri M, Mirzaiyan M, Amini A. Rutting resistance and fatigue behavior of gilsonite-modified asphalt binders. Journal of Materials in Civil Engineering. 2018;**30**:04018292. DOI: 10.1061/(ASCE)MT.1943-5533.0002468

[40] Abraham H. Asphalts and Allied Substances: Their Occurrence, Modes of Production, Uses in the Arts and Methods of Testing. New York, USA: David Van Nostrand; 1918

[41] Helms JR, Kong X, Salmon E, Hatcher PG, Schmidt-Rohr K, Mao J. Structural characterization of gilsonite bitumen by advanced nuclear magnetic resonance spectroscopy and ultrahigh resolution mass spectrometry revealing pyrrolic and aromatic rings substituted with aliphatic chains. Organic Geochemistry. 2012;**44**:21. DOI: 10.1016/j.orggeochem.2011.12.001

[42] Sugihara JM, McCullough TF. Analysis of hydrocarbon fraction of Gilsonite. Analytical Chemistry. 1956;**28**:370. DOI: 10.1021/ac60111a022

[43] Sugihara JM, Sorensen DP. Some pyrroles, phenols and pyridines contained in a Gilsonite distillate. Journal of the American Chemical Society. 1955;**77**:963. DOI: 10.1021/ja01609a049

[44] Soleiman-Beigi M, Noroozian Z, Sarai R, Kohzadi H, Naghipour A. Palladium and zirconium nanoparticles immobilized on functionalized natural asphalt sulfonate as magnetically and

recoverable nanocatalysts for the synthesis of symmetrical and unsymmetrical disulfides. Reaction Kinetics, Mechanisms and Catalysis. 2023;**136**:2465-2480. DOI: 10.1007/s11144-023-02487-9

[45] Bahia BU, Anderson AD. Strategic highway research program binder rheological parameters: Background and comparison with conventional properties. Transportation research record. 2005;**1488**:32-39. DOI: worldcat.org/issn/03611981

[46] Cost-effective and adds a new dimension of strength to asphalt. Americangilsonite company. Available from: https://www.americangilsonite.com

[47] Liu J. Screening Test of Gilsonite Application. Fairbanks, Alaska, USA: Alaska University Transportation Center; 2007

[48] Hawesah AL, Sadique H, Harris M, Al Nageim C, Stopp H, Pearl K, et al. A review on improving asphalt pavement service life using gilsonite-modified bitumen. Sustainability. 2021;**13**:6634. DOI: 10.3390/su13126634

[49] Lv S, Fan X, Yao H, You L, You Z, Fan Z. GAnalysis of performance and mechanism of Buton rock asphalt modified asphalt. Journal of Applied Polymer Science. 2019;**136**:46903. DOI: 10.1002/app.46903

[50] Coleman SR. Structural fat grafts: The ideal filler? Clinics in plastic surgery. 2001;**28**:111-119. DOI: 10.1016/S0094-1298(20)32343-9

[51] Wong C, Ho MK. Effect of Gilsonite-Modified Asphalt on Hot-Mix Asphaltic Concrete Mixes Used in District 12, Houston, Texas. Austin, TX (USA): Texas Dept. of Highways and Public Transportation; 1990

[52] Mixing gilsonite in to bitumen or asphalt mix. Hillinton international group. 2015. Available from: www.gilsonitesuppliers.co.uk

Chapter 7

Road Asphalt Pavements Made by Bottom Ashes from Waste-to-Energy Plant as a Filler

Rosa Veropalumbo, Simone Malvezzi, Cristina Oreto, Francesca Russo and Gianluca Dell'Acqua

Abstract

Not all waste is completely burned by waste-to-energy plants. About 73% of it is bottom ashes (BA). Sustainable reuse of the BA is to reprocess them into the production chain of the asphalt mixtures for road pavements. This research deals with an experimental investigation of the mechanical performance of hot and cold asphalt mastics made with the BA filler versus the traditional ones containing limestone powder (LP). First, environmental compatibility tests were carried out and then a rheological analysis, applying viscosity, frequency sweep, and multiple stress creep and recovery tests, was performed to check the feasibility of using the BA as a filler for asphalt-based materials. Looking at the main results, the ecotoxicological tests in terms of EC50 for toxicity on Daphnia and EC50 for algal growth inhibition confirmed the usability of all the mastics from an environmental point of view. The rheological results showed higher G^* values for cold asphalt mastics made up with the BA compared to the corresponding solutions made with the LP. Additionally, it was also observed that the mastics containing BA returned a 90% decrease in non-recoverable creep compliance at test temperatures over 40°C compared to the corresponding mastics made with the LP.

Keywords: asphalt mixtures, bottom ashes, rheological investigation, ecotoxicity analysis, leaching test

1. Introduction

Potential environmental and mechanical benefits arise from the reuse of some different waste (e.g. plastic, construction and demolition, glass, rubber) for road construction and/or maintenance. One of the main topics today is to mitigate climate change effects associated with the construction of asphalt pavements; in addition, a rational design of pavement superstructures must be followed integrating the mechanical performance with social-economic and sustainable aspects. The application of life cycle assessment methodology can help to highlight the real benefits of each solution beyond the technical performance required [1].

The environmental-socio-economic impact and properties of asphalt paving solutions are very important topics to address at all levels and dimensions; in particular, its 360° integration into the design and maintenance procedures of road pavements is fundamental, together with the other components to improve their quality. An example is the evaluation of the life cycle of all the elements of the road pavement system whose effectiveness is to evaluate the environmental performance of the entire system without compromising the performance [2].

Russo et al. [3] carried out preliminary and base investigations on road asphalt mastics features made up of BA from the Waste-to-Energy plant through laboratory tests. Mechanical performance demonstrated how the rutting resistance at temperatures higher than 40°C increases for alternative mastics compared with the corresponding limestone ones as well as in terms of the elastic behavior that above 30°C improves for cold mastics than the remaining solutions. A leaching test and an X-ray fluorescence analysis were performed for each filler as well as ecotoxicological tests in terms of daphnia EC50 toxicity analysis and an EC50 algal growth test that confirmed the goodness of the environmental sustainability of all investigated solutions.

Many researchers have focused on BA reusing by changing quantity as alternative aggregates for making asphalt mixtures. Among the different topics addressed, there is the examination of some important properties such as the rutting, the low-temperature splitting, the freeze–thaw splitting, and Marshall immersion [4]. The results of these researches disclosed that at high-temperature a mixture made up of 10% of BA over the total weight of the aggregates obtains excellent stability; as concerned the low-temperature crack resistance has been observed as the optimum value when the 20% BA content in the entire asphalt mixture. Other features [5] have concerned the Indirect Tensile Strength (ITS) variation that achieved optimal value for BA close to 6% and as an additional result, an improvement of the skid resistance of all asphalt mixtures.

Many new discoveries regarding the introduction of alternative materials were achieved, above all a perceptible environmental compatibility of asphalt mastics made up by reusing waste from civil engineering field activities showing immobilization of waste into road pavements and no pollutant release exists [6]; in particular, by the study conducted on sustainable mastics made up by filler over neat 50/70 bitumen ratio equal to 1, an average 50% increase of the complex shear modulus was observed compared to the traditional asphalt mastic. In addition, it also guaranteed 35% lower non-recoverable creep compliance values in comparison with the conventional asphalt mastic.

Ing et al. [7] studied for example the effects of four BA amounts on the total weight of the aggregates (0, 2, 4, and 6%) for making hot asphalt mixtures containing 60/70 penetration grade bitumen: Marshall stability, ITS, resilient modulus, and dynamic creep were investigated. The results obtained showed that, for BA content ranging from 2 to 4%, there are better outcomes for the density, stability, stiffness, and flow in comparison to a traditional asphalt mixture without the BA; instead, in correspondence with BA content equal to 6% there is a simultaneous improvement of the ITS, dynamic creep and resilient modulus.

A proper and robust methodological-laboratory structure was carried out by many researchers for carefully investigating and validating the effects of some waste reusing from civil engineering activities as filler for hot and cold asphalt mixtures [8–10]. The introduction of raw secondary materials requires that sustainable asphalt solutions must be properly designed and tested by much more complex laboratory tests and measures than those required for conventional asphalt mixtures. Physical and mechanical properties (i.e., ITS, curing effects for cold recycling, stiffness, rutting, and fatigue resistance) are to be examined. The morphology of each mixture can affect all mixture behavior and it deserves to be studied before moving to test advanced mechanical performance, for example by scanning electron microscopy, XRF, and DSC. The results of many studies have shown that by using a combined amount of waste for the production of cold alternative mixtures it is possible not only to meet the technical standards requirements but also to ensure good mixture performance in addition to the preservation of natural resources.

To date, cold reusing of waste into asphalt mixture has been gaining increasing prominence both for minimizing water-soil contamination and transportation costs and at the same time for reducing the carbon footprint of asphalt mixtures. Nevertheless, cold recycling of raw secondary materials for making asphalt pavements is still a new technique. The evidence of 100% recycled pavement is often a laboratory-methodological investigation, and National Standards and guidelines are missing. Cold recycling of asphalt pavements and some specific raw secondary materials does not negatively impact environmental pollution and human health, also ensuring good mechanical performance [11, 12].

Thus, the purpose of the present research is to assess the technical feasibility of using the BA in cold and hot asphalt mastics through a rheological investigation. Mastics containing traditional limestone powder (LP) filler were mixed for comparison. The filler-over-bitumen ratio (f/b) equal to 0.6, typical of a traditional asphalt mixture, was chosen. Before making mastics, a leaching test was carried out as well and an ecotoxicity investigation was made in terms of the Daphnia toxicity test and the algal growth inhibition test to assess the environmental feasibility of the solutions. Dynamic viscosity, the frequency sweep test, and the multiple stress creep and recovery test were carried out to ascertain the technical feasibility of the mastics with the BA.

2. Material and methods

2.1 Binders

A neat bitumen (NB) 50/70 penetration grade for hot asphalt mastics and a bituminous emulsion (BE) 60/40 for making cold solutions were used (see their properties in **Table 1**).

2.2 Fillers: Limestone powder and bottom ashes

Two fillers were used to blend asphalt mastics: limestone powder (LP) and bottom ashes (BA) (see **Figure 1a**).

The LP originated from the crushing and sieving of limestone rocks in Southern Italy, instead, the BA derives from the combustion process of solid non-recyclable waste, into a Waste-to-Energy plant. The raw BA was first kept in the oven at 100°C for 3 days until to remove the complete moisture and then cleaned from all the ferrous, glassy, and other unburnt elements; after that, the particles deriving from the BA were subjected to a sieving process through the 0.063 mm sieve size to get the filler.

Properties	Unit	Standard	NB	BE
Penetration at 25°C	dmm	EN 1426	68	—
Softening point (R&B)	°C	EN 1427	46	—
Dynamic viscosity at 135°C	Pa•s	EN 13702	0.25	—
Dynamic viscosity at 100°C			4.1	—
Dynamic viscosity at 60°C			185	—
Water content	%	EN 1428	—	40
pH value	—	EN 12850	—	4.2
Settling tendency at 7 days	%	EN 12847	—	5.8

Table 1.
Binders basic properties.

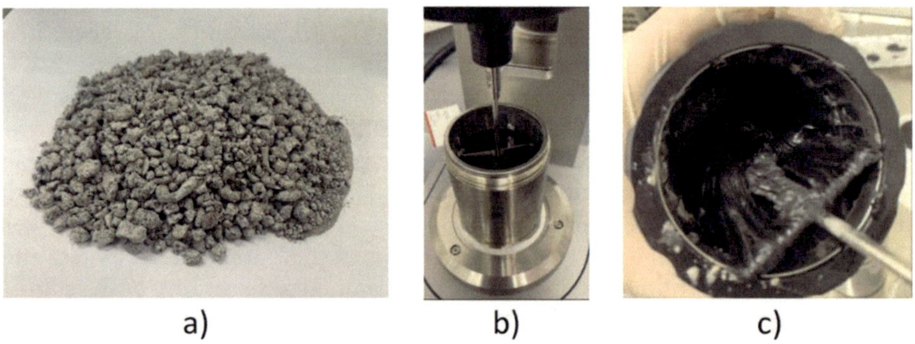

Figure 1.
(a) Bottom ashes, (b) blending of the cold mastic, and (c) cold mastic at the end of the mixing phase.

The LP and BA fillers were both subjected to a leaching test according to the EN 12457-2:2004 standard by using plasma atomic emission spectrometry to detect the presence of the heavy metals before the mastics mixing process [13]. **Table 2** shows the results of the leaching test, which are fully compliant with the limits imposed by the Italian Ministerial Decree of 27 September 2010 for acceptance into inert landfills. The leaching results for both fillers met the mentioned requirements.

2.3 Sample preparation

The mastic samples were prepared by setting the f/b mass ratio to 0.6 in compliance with the most common f/b ratio adopted for asphalt mixture production [14]; additionally, greater f/b ratios were not feasible as the blending of asphalt mastics became considerably difficult due to the mastic stiffening. Aiming to obtain homogeneous samples, the blending procedure was different for the hot and cold production processes.

The hot mixing procedure consisted of three main steps. First, the bitumen and the filler were kept in the oven for 2 h at 160°C to ensure a complete and homogeneous heating of the binder and a free-from moisture filler. After that, the bitumen was poured into an aluminum container previously heated at the same temperature and

Chemical elements	Units	LP	BA	Limits sets by Table 5a of M.D.27/9/2010
DOC	mg/l	11.1	11.2	80
Total dissolved solids (TDS)	mg/l	607	1760	6000
Chloride	mg/l	2.34	80	1500
Fluorides	mg/l	3.32	0.0168	15
Sulfate	mg/l	91	27.9	2000
Antimony	mg/l	0.00178	0.0261	0.07
Arsenic	mg/l	0.00131	n.d.	0.2
Barium	mg/l	0.029	0.27	10
Cadmium	mg/l	n.d.	n.d.	0.1
Chromium	mg/l	0.0111	0.059	1
Mercury	mg/l	0.000121	0.000112	0.02
Molybdenum	mg/l	0.00762	0.068	1
Lead	mg/l	n.d.	0.0178	1
Nickel	mg/l	n.d.	n.d.	1
Copper	mg/l	0.00408	0.046	5
Selenium	mg/l	0.00333	0.00164	0.05
Zinc	mg/l	0.00224	0.0421	5

Table 2.
Leaching test results.

then the right amount of filler previously weighted was progressively added. A propeller impeller at 4000 rpm for 10 min was used to mix the whole compound. Finally, the obtained hot solutions were stored in an aluminum container from which samples were extracted for the execution of the tests described below.

Concerning the cold mastics, a particular mixing system named building material cell (BMC) (see **Figure 1b**) was used to blend the asphalt mastics; the impeller consists of a shovel shape, and the mixing speed and temperature are controlled by the dynamic shear rheometer (DSR) Anton Paar smart pave 102. The binder used in this case was the BE that was previously conditioned in the oven at 60°C for 60min; at the same time, the filler was kept in the oven at 60°C for 30 minutes to guarantee a homogeneous mixing temperature. The right amount of BE was poured into a BMC system, also heated at 60°C, and then positioned on the Peltier system of the DSR, previously set to 60°C. At that point, the shovel impeller was lowered into the BMC container. The DSR was set at a shear rate of 30 1/s, and the BE was stirred for 1 min. After 1 min, the amount of previously weighed and heated filler was gradually added, until complete mixing after 20 min (see **Figure 1c**). The viscosity of the solutions was constantly checked throughout the whole process in order to identify the breaking of the bitumen emulsion i.e., the instant when the water phase separates from the bitumen, creating the cold asphalt mastic.

Table 3 shows the list of the asphalt mastic solutions object of the investigation.

Solution ID	Binder type	Fille type	Mixing process
HML	NB	LP	Hot
HMB	NB	BA	
CML	BE	LP	Cold
CMB	BE	BA	

Table 3.
Asphalt mastics identification.

2.4 Ecotoxicity tests

The ecotoxicity test is a biological experiment aimed at verifying whether a potentially toxic compound causes a relevant biological response in the organisms used for the test.

The ecotoxicity of the organism *Daphnia magna* crustacean (UNI EN ISO 6341: 2012) and growth inhibition of Pseudochirkneriella subcapitata green algae (EN ISO 8692: 2012) were assessed according to the Regulation (EC) 440/2008 and the OECD Guidelines.

The EC50, namely the concentration with toxic effect for 50% of organisms, was determined for a solution with a concentration equal to 100 mg/L, stirred at 100 rpm with an orbital stirrer. The contact time between the sample and the liquid was 7 days. Then, the solution was decanted. The aqueous fraction of the solution was taken and used for the exposure.

For the algal growth inhibition test, several concentrations containing the solutions to be tested (39, 63, 100, 160, 256 mg/L) versus the negative control were inoculated with a known volume of culture in exponential growth in order to have an initial cell concentration of 10,000 cells/mL. The flasks were incubated and kept in agitation for a period of 72 h (see **Figure 2a**). At the end of the exposure period, cell density was measured using an electronic cell counter. Growth inhibition was determined in comparison to the control culture.

For the acute toxicity test on *Daphnia magna*, four replicates containing the solutions were to be tested for each concentration and four negative controls were prepared. In each replica, 5 daphnids were introduced aged less than 24 h. The beakers were incubated under controlled conditions inside a climatic chamber (see **Figure 2b**). After 24 and 48 h from the beginning of the exposure, immobilization was recorded and the results were analyzed to calculate the EC50 value.

2.5 Delta ring and ball

The change in stiffening of the binder due to the addition of the filler was evaluated by means of the measurement of the delta ring and ball value (ΔR&B). ΔR&B is calculated as the difference between the softening point of the mastics and the softening point of the bitumen used for its production, each one determined in compliance with the EN 1427 standard.

2.6 Rheological properties tests

The rheological investigation of the solutions prepared and evaluated within the present study was carried out by using the DSR previously mentioned.

a) b)

Figure 2.
Ecotoxicity tests: (a) oscillating plane, and (b) climatic chamber.

The cone-plate method with a cone of 25 mm diameter and 1° angle was adopted to measure the dynamic viscosity in accordance with the EN 13702 standard. The test was carried out at 60°C temperature and applied a shear rate of $5 \times 10^{-2}\ s^{-1}$.

The dynamic viscosity was calculated as follows:

$$\eta = \tau / \dot{\gamma} \tag{1}$$

where

- $\dot{\gamma}$ is the shear rate expressed in s^{-1}
- τ is the stress expressed in Pa

The complex shear modulus (G*) and the phase angle (δ) were measured by performing the frequency sweep (FS) test. The FS was conducted using a plate-plate system with a 25 mm diameter and 1 mm gap from −10 to 20°C temperature, while a plate-plate system with 8 mm diameter and 2 mm gap was adopted for temperatures ranging from 30 to 60°C. For both the temperature range, nineteen frequencies starting from 0.1 until 10 Hz were analyzed. The viscoelastic linear range was determined in compliance with Annex C of the EN 14770 standard before carrying out the FS test. In particular, for the 25 mm plate-plate geometry, the strain sweep was carried out at the highest temperature and lowest frequency; vice-versa, for the 8 mm plate-plate geometry, the last strain value at which the G' and G" components do not differ by more than 5% of the initial value was detected. Strains equal to 0.01 and 0.5% were obtained respectively for the 8 mm and 25 mm geometries.

The evaluation of the elastic response of the prepared solutions was carried out through the multiple stress creep and recovery (MSCR) tests at 0.1 and 3.2 kPa stress levels. At each stress level, 1 s creep and 9 s recovery phases were applied. The test temperatures were equal to 40, 50, and 60°C. Aiming to express the resistance of the solutions to permanent deformation under repeated load and their elasticity, respectively the non-recoverable creep compliance (Jnr) and the percentage of recovery (R%) were calculated by using the following equations:

$$\%R_{\tau} = \frac{1}{10}\sum_{N=1}^{10}\left[100\cdot\left(\varepsilon_1^N - \varepsilon_{10}^N\right)/\,\varepsilon_1^N\right] \tag{2}$$

$$Jnr_{\tau} = \frac{1}{10}\sum_{N=1}^{10}\left(\varepsilon_{10}^N / \tau\right)\ \left(kPa^{-1}\right) \tag{3}$$

where

- ε_1^N is the total adjusted strain value at the end of the creep portion (ie., after 1 s) of each N-th cycle. The index N stands for the number of cycles since there is an ε_1 for every cycle
- ε_{10}^N is the non-recovered adjusted strain value at the end of the recovery portion (that is, after 10 s) of each N-th cycle
- τ is the applied creep stress level considered: 0.1 and 3.2 kPa

3. Results

3.1 Ecotoxicity analysis

Ecotoxicity tests have been used to evaluate the toxic effects of chemical and physical agents on living organisms.

In the present research two different types of tests were performed, i.e. the EC50 Daphnia toxicity test and the EC50 algal growth test.

The ecotoxicity test both presented an EC50 value greater than 250 mg/L, confirming that the solutions designed in the present work are not toxic for the aquatic environment.

3.2 Stiffening effect

The delta ring and ball test results are presented in **Figure 3**. It is noticeable that the cold mastics present a greater value of ΔR&B than the hot ones; in particular, CMB compared to HMB resulted in 7.1°C greater ΔR&B, and CML 9.8°C greater ΔR&B than HML. This could be due to a better dispersion of the filler inside the cold mastics during the mixing phase so as to create a more solid filler matrix within it. Anyway, the BA presents higher ΔR&B values compared to the solutions containing the LP, about 7.7°C higher in the case of a hot mixing process and 5°C higher when added cold. These results show that the addition of BA benefits the mastic in terms of stiffness much more than a traditional filler, underlining a greater advantage from the point of view of resistance to accumulation of deformation. In brief, the CMB solution shows the highest stiffening effect among all the mastics analyzed.

3.3 Dynamic viscosity

The results of the dynamic viscosity at 60°C are illustrated in **Figure 4**. As can be seen from the comparison of the binders (NB and BE) and mastic solutions (HML,

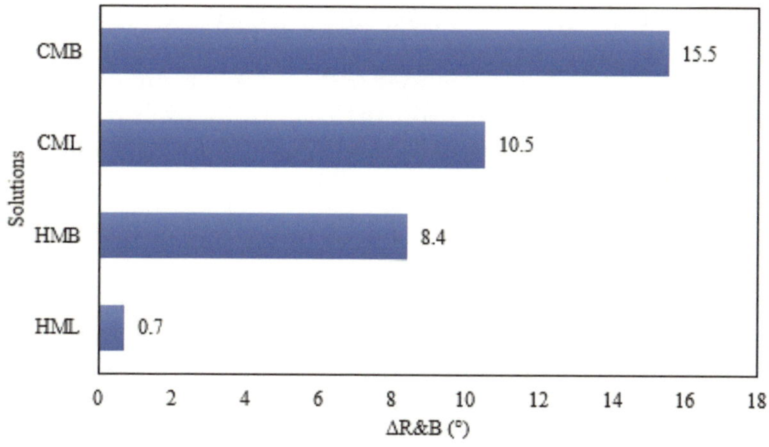

Figure 3.
Delta ring and ball results.

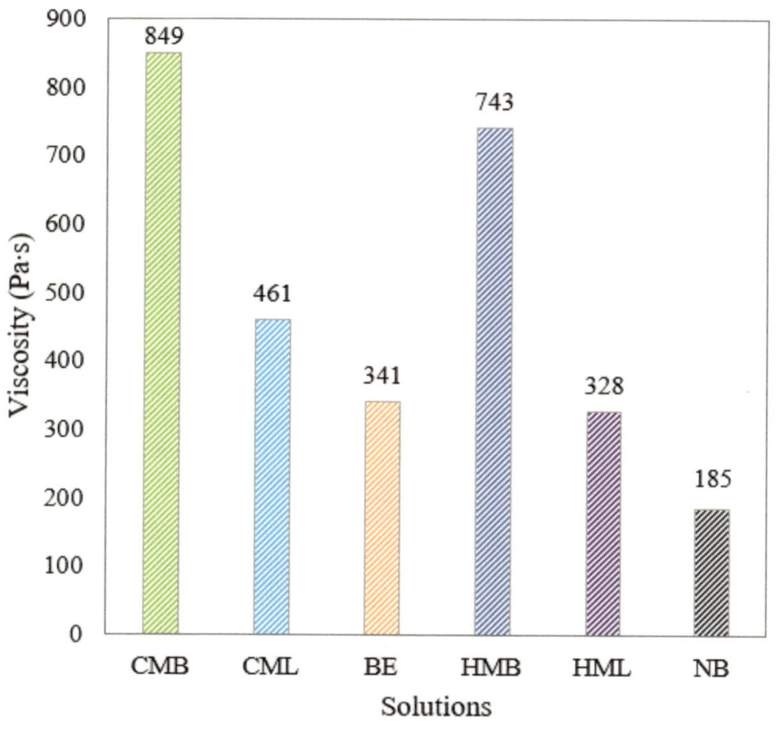

Figure 4.
Viscosity at 60°C of the asphalt maatics and binder solutions.

HMB, CML, and CMB), higher viscosity values result for the latter; in particular, the viscosity of the binder increases on average by 56% with the addition of LP while it increases on average by 225% with the addition of BA.

Usually, the measurement of the viscosity at 60°C is useful to predict the behavior of asphalt binders under high pavement service temperatures [15], with specific attention to permanent deformation: high viscosity values imply a good resistance to the accumulation of permanent deformations. Among the mastics analyzed, it appears that when the fillers are added cold there is a significant increase in viscosity compared to mastics produced according to the hot mixing process; in fact, the viscosity of CMB resulted in 14% higher than that of HMB, while that of CML is 41% higher than HML. Nevertheless, the addition of BA always brings a more beneficial effect than LP when added to produce either cold or hot asphalt mastic solutions.

3.4 Frequency sweep

In order to explain the results of the frequency sweep test in a wide range of frequencies leaving aside the temperature dependence, the master curves were designed. In general, a master curve is created by shifting the curves from tests at different temperatures to generate a single smooth curve at a reference temperature [16]. The Williams-Landel-Ferry (WLF) principle [17] was adopted in the present study.

The master curves of the four mastics (HML, HMB, CML, and CMB) and two binders (NB and BE) are presented in **Figure 5**. Firstly, it can be observed that all the mastics exceed the binders in terms of G* in the whole frequency range, in particular, HML and HMB exceed by more than 50% NB, and CML and CMB exceed by more than 80% EB.

Above the reduced frequency of 80 Hz, CML superimposes HMB resulting in the mastics with the highest G* values followed by HML and CMB, which in turn return the same values of G*. This behavior denotes that, for high values of reduced frequencies, corresponding to low temperatures, the addition of BA filler affects much more the hot mastics much as it produces higher stiffness values, contrary to what occurs with the LP that benefits more the cold mastics.

As concerns the phase angle above 80 Hz, the solutions containing BA (HMB and CMB) returned lower values, about 50% than the mastics made up of LP (HML and CML) and the two binders (NB and EB).

Considering the frequencies lower than 80 Hz, it is possible to notice a change in the behavior of the CMB which tends to override all the other mastic solutions showing higher G* values. At the same time, the trend of HMB and CML remains constant. In correspondence with the lowest reduced frequencies, CMB reached a G*

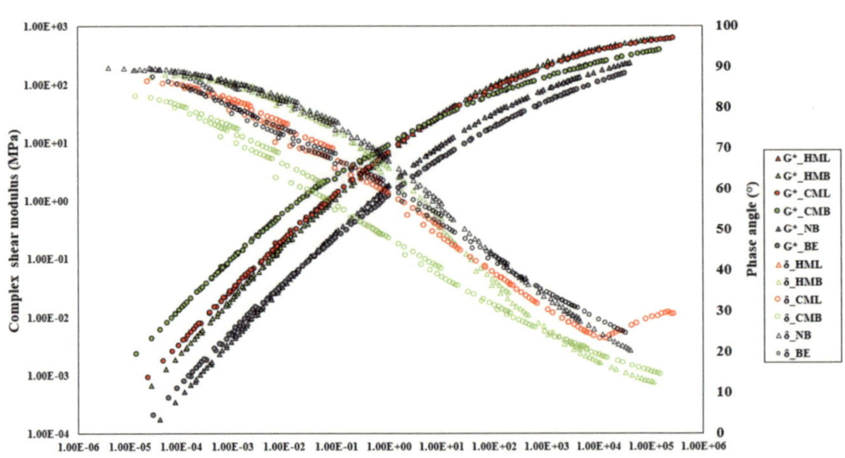

Figure 5.
Master curves shoving G vs. reduced frequency and phase angle vs. reduced frequency.*

value equal to 2.28E-03 MPa, which is 156, 245, and 551% higher than CML, HMB, and HML, respectively.

Also, in terms of phase angle, the CMB solutions always give the lowest values compared to all the other solutions, starting from 25° at around 80 Hz until 82° at 1.35E-05 Hz.

The results above described revealed that CMB is an excellent solution for resisting low-frequency loads, given the highest stiffness; anyway, good responses have been also obtained by the hot mastics with BA. The greater stiffness of cold mastics with BA could be due to the chemical interactions between the BA and water contained in the bituminous emulsion, which may result in a cement-like hardening process. While the LP is a non-reactive filler, the BA originated from feedstock that may include different materials including construction residues and waste. As a matter of fact, the BA goes through a carbonatation reaction that begins after the BA is produced during the waste-to-energy process and mostly ends after 2 to 3 weeks; the reactivity of the BA with water may be also linked to a residual carbonatation activity.

For an easy visualization of the relation between the phase angle and G*, regardless of temperature and frequency, a black diagram was plotted (see **Figure 6**). The black diagram is a useful tool to compare the elastic behavior between different solutions having fixed the stiffness parameter, in this case, the G*. Looking at **Figure 6**, different elastic response is revealed from the comparison of the solutions. At δ below 30°, all the solutions do not show a significant difference among them in terms of elasticity, having fixed the value of G*. In this case, the only difference that can be noticed refers to the solutions HMB and CMB: both showed the highest G* values in correspondence with the lowest phase angle value. Anyway, the CMB compared to the other mastic solutions (HMB, HML, and CML) always has a lower phase angle at the same G*; on the contrary, CML always has a higher phase angle value, fixed the G*.

When the phase angle tends to increase (see **Figure 6**), the difference among the asphalt mastic solutions begins to grow: until a phase angle value equal to 60°, the BE and CMB always have a lower phase angle, about 70% difference than the remaining solutions (NB, HMB, HML, and CMB). In this latter case, HML and CML provide the same response with the highest phase angle values.

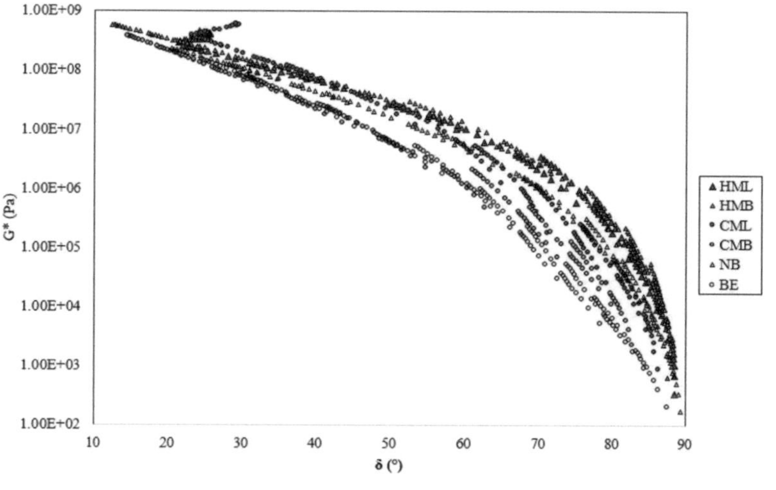

Figure 6.
Black diagram: G vs. phase angle.*

Above δ equal to 60°, CML and CMB appear as mastics with the lowest phase angle value, never reaching the highest phase angle value, as well as BE whereas it shows the absolute lowest phase angle value in correspondence with the lowest G*.

Noticed the lowest and maximum values of phase angle that the cold mastic solution made up of BE can reach, given the high G* values, CMB represents a good solution in terms of the resistance to permanent deformation accumulation, as previously demonstrated by the viscosity analysis. Indeed, rutting and viscosity are well known to be strongly correlated as previously shown in the literature [18].

3.5 MSCR results

Figure 7 presents the results of MSCR test in terms of Jnr of the four asphalt mastics (HML, HMB, CML, and CMB) and two binder solutions (NB and BE). In more detail, **Figure 7a** shows the results at 0.1 kPa stress level at all three test temperatures analyzed (40, 50, and 60°C). It can be immediately noticed that, as the temperature increases, starting from 40 until 60°C, the Jnr values of all the solutions increase; in particular, at 40°C the Jnr values are on average 82% lower than those at 50°C, which in turn are 76% lower than the Jnr values calculated at 60°C.

Comparing the mastic solutions with the respective binder used for the production at 40°C resulted in the Jnr values of HML and HMB being respectively 52 and 81% lower than NB, while those of CML and CMB were 86 and 96% lower than EB. At the same temperature of 40°C, fixing the type of mixing, the Jnr value of HMB is 61% lower than HML and that of CMB is 73% lower than CML.

These results demonstrate how the addition of filler in bitumen improves its resistance properties to the accumulation of permanent deformations, especially when BA is added with the cold mixing process.

When the test temperature increases, the difference between mastics containing LP and those containing BA remains almost the same moving from 40 to 50°C and from 50 to 60°C. In particular, the Jnr value showed by CMB is always lower than that of all other mastics (CML, HML, and HMB), on average 91% lower at 50°C and 88% lower at 60°C.

Figure 7b illustrates the Jnr values measured at 3.2 kPa, where it can be observed that also in this case when the temperature increases the Jnr value increases in the same measure as the stress level of 0.1 kPa.

From the comparison with the bitumen, the CMB solutions present the highest difference, in particular, the Jnr value of CMB is on average 95% lower than that of EB at all test temperatures (40, 50, and 60°C).

Comparing the effect of BA with LP, it resulted that at all test temperatures, the Jnr value of HMB is on average 61% lower than that of HML and that of CMB is on average 65% lower than that of CML.

It is interesting to observe the difference in terms of Jnr moving from 0.1 to 3.2 kPa: no substantial increment of Jnr has been revealed. In fact, the highest Jnr increase of 32% is obtained by CML. It can be seen how mastics containing BA react well to high test temperatures, from 0.1 to 3.2 kPa, at 60°C, the Jnr of HMB increases only by 7% and that of CMB just by 5%.

These results state that the mastics are not so stress-dependent as the binders, in particular at high test temperatures, where it is expected that the accumulation of the permanent deformation grows more sharply, the solutions designed in the present research behave in the same way in 0.1 and 3.2 kPa.

To denote the elastic nature of the asphalt mastic solutions under exam after the unloading cycle, the %R was calculated. Usually, a high value of %R indicates that

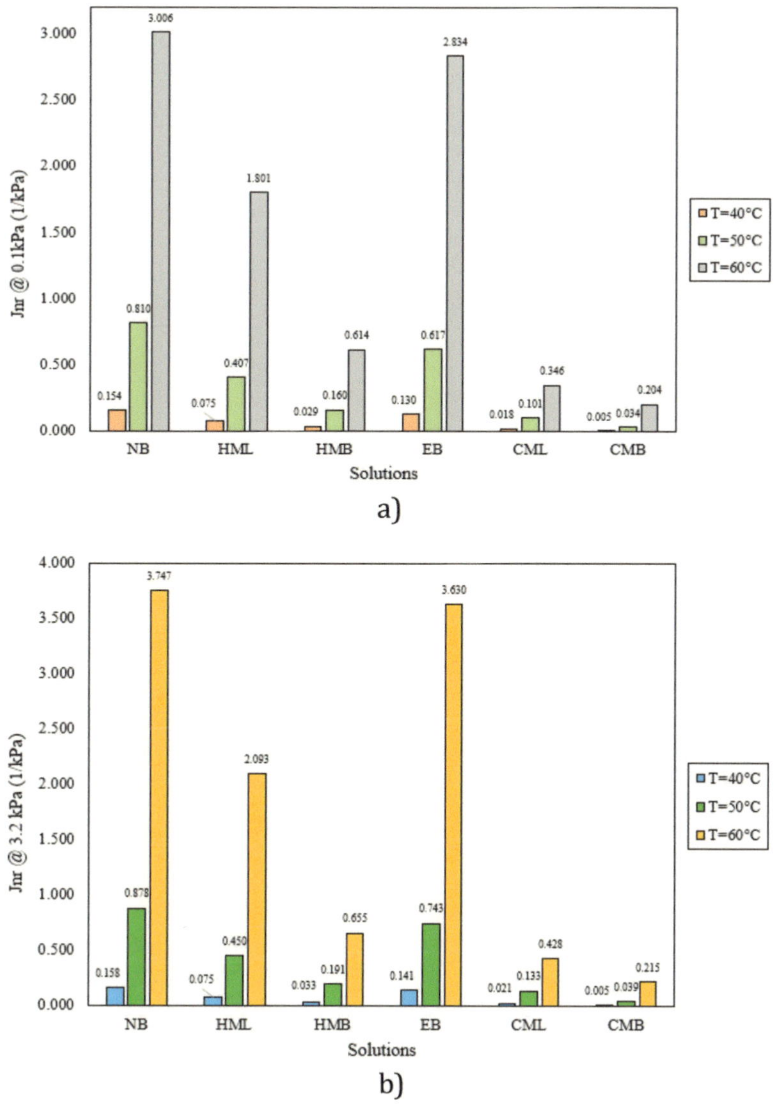

Figure 7.
MSCR results at 40, 50 and 60°C: (a) Jnr @0.1 kPa, and (b) Jnr @3.2 kPa.

the solution recovers very quickly after load cycles therefore minimizing permanent deformations [19]. The results of the %R of each solution as a function of the temperature and stress level are reported in **Table 4**.

By looking at the comparison between the asphalt mastics and each binder used for the mixing, different behaviors have been shown at the temperature of 40°C, regardless of the stress level: in the case of the hot mastics, HML gives on average 8% lower %R values than NB, whereas HMB shows on average 24% higher %R values than NB.

Concerning the cold mastics under 3.2 kPa stress level, the %R value of CML resulted in 13% lower than that of EB, differently from the CMB solution that presented 97% higher %R value than that of EB.

At test temperatures of 50 and 60°C, all the solutions present at both stress levels had higher %R in comparison to the original binder.

Solutions	%R @40 °C (1/kPa)		%R @50 °C (1/kPa)		%R @60°C (1/kPa)	
	0.1 kPa	3.2 kPa	0.1 kPa	3.2 kPa	0.1 kPa	3.2 kPa
NB	3.377	2.700	1.191	0.453	0.701	0.235
HML	3.308	2.443	1.546	0.639	0.964	0.260
HMB	4.559	3.052	2.064	0.696	1.145	0.446
EB	9.613	7.318	4.890	1.698	1.897	0.016
CML	14.041	6.399	5.297	1.828	4.093	0.680
CMB	19.075	14.424	9.516	5.133	5.023	1.171

Table 4.
Percentage recovery results.

In addition, each cold mastic, in comparison to the corresponding hot ones, regardless of the temperature and load level, showed higher %R values. Significant is the increment of %R of CMB that is seven times higher than HMB under the stress level of 3.2 kPa at a temperature of 50°C.

Looking at the comparison between CML and CMB, the latter returned always higher %R values; in fact, averaging both stress levels, at 40, 50, and 60°C CMB has a %R respectively 81, 130, and 47% higher than that shown by CML.

The obtained results are complementary to what was previously demonstrated with the discussion of Jnr values, i.e., that the addition of BA as filler lowers the sensitivity of both hot and cold mastics to the accumulation of permanent deformations since it provides a greater degree of elasticity.

4. Conclusions

The research evaluated the feasibility of treating the filler from bottom ashes as a possible substitute for a traditional limestone powder filler for asphalt mixture production; the application of a new technology, where the mixing temperatures are reduced from 160 to 60°C, was applied to produce asphalt mastics. The filler-over-bitumen ratio equal to 0.6 was used to prepare four different asphalt mastics made with the cold and hot mixing process, both with bottom ashes and limestone powder fillers. The leaching test results confirmed that the fillers are not hazardous and the ecotoxicity analysis carried out on the packaged mastics solutions revealed the non-release of toxic substances into the aquatic environment.

From the empirical and rheological analysis carried out through the frequency sweep test and multiple stress creep and recovery test, the following conclusions can be drawn:

- A greater stiffening effect was observed for the mastics containing bottom ashes as filler since a higher value of the delta ring and ball emerged from the comparison with the mastics containing limestone powder; indeed, the latter had a 7°C lower delta ring and ball value than the mastics with bottom ashes.

- At lower reduced frequencies the cold mastics solutions behave generally better than all the others analyzed in the present research, showing the highest complex shear modulus values; the cold solution made up of bottom ashes presented a shear modulus value 156% higher than the cold solution containing limestone powder filler.

- When the fillers are added to the binder, whether hot or cold, they improve the resistance to permanent deformation accumulation, returning lower non-recoverable creep compliance as that of the hot solution made up of bottom ashes is 61% lower than the traditional hot mastic made up of limestone powder filler.

- The multiple stress analysis revealed that the designed mastics are not stress-dependent; as a matter of fact, no significant changes were appreciated in terms of accumulation of deformation moving from 0.1 to 3.2 kPa stress level.

- The addition of bottom ashes into cold mastics produced solutions with a high degree of elasticity, returning at the same time the lowest phase angle and the lowest values of percentage recovery in correspondence with the highest test temperature, equal to 60°C.

All the above-mentioned benefits are expected to reflect on the durability, rutting susceptibility, and overall regularity of the asphalt pavement where the asphalt mastic is made of BA instead of traditional limestone filler. In order to ascertain the increase in the service life and reduction of the ordinary maintenance frequency of such pavements, future studies will be carried out to mix BA mastics with coarse aggregates to produce the actual asphalt mixture and investigate its susceptibility to rutting.

Acknowledgements

This research was supported by A2A Ambiente SpA.

Conflict of interest

The authors declare no conflict of interest.

Appendices and nomenclature

BA	Bottom ashes
FA	Fly ashes
f/b	Filler-over-bitumen ratio
NB	Neat bitumen
BE	Bituminous emulsion
LP	Limestone powder
BMC	Building material cell
DSR	Dynamic shear rheometer
ΔR&B	Delta ring and ball value
G*	Complex shear modulus
δ	Phase angle
Jnr	Non-recoverable creep compliance
%R	Percentage of recovery

Author details

Rosa Veropalumbo[1], Simone Malvezzi[2], Cristina Oreto[1], Francesca Russo[1*] and Gianluca Dell'Acqua[1]

1 University of Naples Federico II, Naples, Italy

2 A2A Life Company, Brescia, Italy

*Address all correspondence to: francesca.russo2@unina.it

References

[1] Russo F, Veropalumbo R, Oreto C. Climate change mitigation investigating asphalt pavement solutions made up of plastomeric compounds. Resources, Conservation and Recycling. 2023;**189**:106772. DOI: 10.1016/j.resconrec.2022.106772

[2] Russo F, Oreto C, Veropalumbo R. Promoting resource conservation in road flexible pavement using jet grouting and plastic waste as filler. Resources, Conservation and Recycling. 2022;**187**:106633. DOI: 10.1016/j.resconrec.2022.106633

[3] Russo F, Veropalumbo R, Oreto C, Cassese D, Papa B, Malvezzi S. Reusing bottom ash as a filler from a waste-to-energy plant for making asphalt mastics. Case Studies in Construction Materials. 2022;**17**:e01406. DOI: 10.1016/j.cscm.2022.e01406

[4] Ou J, Yan K, Li L, Long Z, Zhou W, Xiao L, et al. Conversion of municipal solid waste incineration bottom ash in asphalt pavements. Advances in Civil Engineering Materials. 2022;**11**(1):138-154. DOI: 10.1520/ACEM20210032

[5] Buritatum A, Suddeepong A, HorpibulsukS, AkkharawongwhatthanaK, Yaowarat T, Hoy M, et al. Improved performance of asphalt concretes using bottom ash as an alternative aggregate. Sustainability. 2022;**14**(12):7033. DOI: 10.3390/su14127033

[6] Russo F, Veropalumbo R, Pontoni L, Oreto C, Biancardo SA, Viscione N, et al. Sustainable asphalt mastics made up recycling waste as filler. Journal of Environmental Management. 2022;**301**:113826. DOI: 10.1016/j.jenvman.2021.113826

[7] Ing NLS, Ming NC, Hasan M, Jaya RP, Shaffie E, Mashros N, et al. The influence of coal bottom ash as filler in asphalt mixture. Key Engineering Materials. 2022;**912**:185-198. DOI: 10.4028/p-beny4x

[8] Russo F, Veropalumbo R, Biancardo SA, Oreto C, Scherillo F, Viscione N. Reusing jet grouting waste as filler for road asphalt mixtures of base layers. Materials. 2021;**14**(122):3200. DOI: 10.3390/ma14123200

[9] Viscione N, Lo Presti D, Veropalumbo R, Oreto C, Biancardo SA, Russo F. Performance-based characterization of recycled polymer modified asphalt mixture. Construction and Building Materials. 2021;**310**:125243. DOI: 10.1016/j.conbuildmat.2021.125243

[10] Veropalumbo R, Viscione N, Russo F. Rheological and Mechanical Properties of HMA Containing Fly Ashes as Alternative Filler. In: Pasetto M, Partl M, Tebaldi G, editors. Proceedings of the 5th International Symposium on Asphalt Pavements & Environment (APE). ISAP APE 2019. Lecture Notes in Civil Engineering. Vol. 48. Cham: Springer; 2020. DOI: 10.1007/978-3-030-29779-4_9

[11] Liu Z, Sun L, Zhai J, Huang W. A review of design methods for cold in-place recycling asphalt mixtures: Design processes, key parameters, and evaluation. Journal of Cleaner Production. 2022;**370**:133530. DOI: 10.1016/j.jclepro.2022.133530

[12] Lokesh G, Rakesh K. Experimental investigation of reclaimed asphalt foamed bituminous mix: A cold mix technique. World Journal of Engineering.

2023;**20**(4):655-6684. DOI: 10.1108/WJE-10-2021-0572

[13] Veropalumbo R, Oreto C, Viscione N, Pirozzi F, Pontoni L, Trancone G, et al. Exploring the effect on the environment of encapsulated micro-and nano-plastics into asphalt mastics for road pavement. Environmental Research. 2023;**216**:114466. DOI: 10.1016/j.envres.2022.114466

[14] Lei B, Xiong Q, Zhao H, Dong W, Tam VW, Sun Z, et al. Performance of asphalt mortar with recycled concrete powder under different filler-to-asphalt weight ratios. Case Studies in Construction Materials. 2023;**18**:e01834. DOI: 10.1016/j.cscm.2023.e01834

[15] Biro S, Gandhi T, Amirkhanian S. Determination of zero shear viscosity of warm asphalt binders. Construction and Building Materials. 2009;**23**(5):2080-2086. DOI: 10.1016/j.conbuildmat.2008.08.015

[16] Zhang J, Luo T, Cui Y, Xie B, Wang J, Yan C. Characterizing ZSV of SBS modified asphalt using shear rate sweep test and frequency sweep oscillation test. Construction and Building Materials. 2023;**371**:130757. DOI: 10.1016/j.conbuildmat.2023.130757

[17] Veropalumbo R, Russo F, Viscione N, Biancardo SA. Rheological properties comparing hot and cold bituminous mastics containing jet grouting waste. Advances in Materials Science and Engineering. 2020:1-16. DOI: 10.1155/2020/8078527

[18] Sybilski D. Relationship between absolute viscosity of polymer-modified bitumens and rutting resistance of pavement. Materials and Structures. 1994;**27**:110-120. DOI: 10.1007/BF02472829

[19] Kataware AV, Singh D. A study on rutting susceptibility of asphalt binders at high stresses using MSCR test. Innovative Infrastructure Solutions. 2017;**2**:1-9. DOI: 10.1007/s41062-017-0051-1